Workplace Communications

Workplace Communications

The Basics

George J. Searles

Mohawk Valley Community College

Allyn and Bacon

Boston ■ London ■ Toronto ■ Sydney ■ Tokyo ■ Singapore

Vice President and Editor-in-Chief, English: Joseph Opiela
Editorial Assistant: Marybeth Varney
Cover Administrator: Jenny Hart
Composition Buyer: Linda Cox
Manufacturing Administrator: Suzanne Lareau
Marketing Manager: Kate Sheehan
Production Coordinator: Deborah Brown
Editorial-Production Service: Susan McNally
Text Design and Electronic Composition: Denise Hoffman

A Viacom Company
160 Gould Street
Needham Heights, MA 02494

Internet: www.abacon.com

Library of Congress Cataloging-in-Publication Data

Searles, George J. (George John), 1944–
Workplace communications—the basics / by George J. Searles.
p. cm.
Includes index.
ISBN 0-205-27121-9
1. English language—Business English. 2. English language—Technical English. 3. Technical writing—Problems, exercises, etc. 4. Business writing—Problems, exercises, etc. I. Title.
PE1479.B87S43 1998
808'.06665—dc21 98-33687
CIP

Printed in the United States of America

10 9 8 7 6 5 4 3 2 1 03 02 01 00 99 98

To

the Gage Family

Contents

Preface

This book was written as the solution to a problem. Semester after semester, I had searched unsuccessfully for a suitable text to use in my English 110 course, entitled "Oral and Written Communication," at Mohawk Valley Community College. Designed as an alternative to traditional first-year composition, the course satisfies curricular English requirements for students anticipating careers in such fields as welding, heating and air conditioning, and case management. As might be expected, English 110 is a highly practical, hands-on course that meets the specialized needs of its target audience by focusing on job-related communications.

Although some excellent texts have been written in the fields of business and technical communications, nearly all are aimed at the university level, and are therefore beyond the scope of a course such as English 110. Finally I decided to fill the gap and met my students' needs by creating a textbook of my own. *Workplace Communications: The Basics* has been more than five years in the making, having undergone extensive fine-tuning. I field-tested a number of preliminary drafts in my classes at Mohawk Valley, and students responded enthusiastically, citing the book's accessibility, clarity, and pragmatic, down-to-earth emphasis as particularly appealing features.

Short on theory, long on practical application, written in a simple, conversational style, *Workplace Communications* is exceptionally user-friendly. The book is appropriate not only for recent high school graduates, but also for adult students returning to college, and other non-traditional learners. It is comprehensive and challenging enough for trade school and community college courses such as English 110, and also for similar introductory level classes at most four-year institutions. Its helpful features include:

- learning objectives and outlines for each chapter
- examples, illustrations, and exercises, based on actual workplace contexts
- helpful checklists at the end of major sections
- integrated discussion of computerized information technology and its impact on contemporary workplace practices

I am really looking forward to using *Workplace Communications: The Basics* in my classroom, and hope that other instructors will find it valuable as well. Please send your comments and suggestions to my e-mail address, g.searles@mvcc.ed, or by conventional mail to the Humanities Department, Mohawk Valley Community College, 1101 Sherman Drive, Utica, New York, 13501.

Acknowledgments are in order. First, I wish to thank my principal academic mentors: Sheldon N. Grebstein, Harold Cantor, and the late David Bernstein. I must also mention my students, for they too have taught me much. As always, I am grateful for the assistance of the MVCC Library staff, especially Sherry Day, Ron Foster, and Mary Beth Portley. Thanks also to my reviewers, Barry Batorsky, DeVry Institute of Technology, North Brunswick, New Jersey; Ray Dumont, University of Massachusetts, Dartmouth; Eddye S. Gallagher, Tarrant County Junior College; Rima Gulshan, University of Maryland, Eastern Shore; Susan Griffiths, Portland Community College; Del McGinnis, Delgado College; Nell Ann Pickett, Hinds Community College; Richard Profozich, Prince George Community College; Sherry Sherrill, Forsyth Technical Community College; Maureen Schmid, DeVry Institute of Technology, Phoenix; Jack Schreve, Allegany Community College; Wendy Slobodnik, Heald Business College; Katherine Staples, Austin Community College; Ann Thomas, San Jacinto Community College; Arthur Wagner, Macomb Community College; and Diane Witmer, DeVry Institute of Technology, Dallas, whose encouragement and constructive criticism of the manuscript were helpful indeed. I am indebted as well to my editor at Allyn & Bacon, Joe Opiela, and to his fine staff: marketing manager Kate Sheehan and assistant Rebecca A. E. Ritchey, copyeditor Julie Collins, production coordinators Deborah Brown and Susan McNally, and designer Denise Hoffman, Glenview Studios.

Most importantly, I thank my wife, Ellis, and my sons, Jonathan and Colin . . . for everything.

Introduction

As even its title suggests, *Workplace Communications: The Basics* is in no sense a typical English textbook.

Appropriate as such topics may be in a traditional composition text, you will find nothing here about how to write 500-word essays, and nothing about how to critique English literature. Although some important fundamentals of correct composition are reviewed in an appendix, the text itself refers very little to formal grammar. Instead, it focuses on such matters as the purpose, audience, and tone of communications. Throughout, there is great emphasis on the essential features of effective workplace writing: concision, clarity, and proper formatting. In keeping with the book's highly practical nature, you will work exclusively with non-academic forms of writing, the kind done on the job. Among these are memos, business letters—including the application letter and résumé—and written reports both short and long.

In addition, chapters detail how to handle specific tasks such as writing summaries, descriptions, and instructions; how to deliver oral reports; and how to enhance your oral and written presentations by using visual aids such as tables and graphs. Every chapter section includes numerous examples and illustrations, along with exercises that enable you to practice applying specific principles. Through the use of common-sense strategies, you will learn to express yourself quickly and directly, with no wasted words. And once you begin to communicate more confidently and efficiently, you will be better motivated to eliminate any basic mechanical errors that have weakened your writing in the past.

The communication skills you will develop are important not simply for the sake of completing a course and satisfying an English requirement. Combined with specialized training in your major field of study, these skills will also help equip you for success in the highly competitive environment of the twenty-first-century workplace. In survey after survey, employers repeatedly mention good communications skills along with character, technical knowledge, and computer literacy when asked what they consider to be the most desirable attributes a job candidate can possess. Fortunately, you need not major in English to learn to communicate better. Anyone can. It requires only three components: desire, effort, and guidance. The first two are your responsibility. Coupled with your instructor's efforts, this text will provide the third.

The content of *Workplace Communications* is based on the author's own workplace experiences gained during more than twenty-five years not only as a writing teacher but also as a professional social worker, widely published freelance journalist, and communications consultant to numerous businesses, organizations, and social services agencies. The emphasis, therefore, is not on abstract theory but on practical application. This text is designed specifically for *you*—the student. After completing the book you will know a great deal more than you did before about written and oral communication in the workplace. You will be better prepared to confront any such challenges your chosen career presents. And if you decide to continue your education, what you have learned will provide a solid foundation for further study.

1

The Keys to Successful Communication: Purpose, Audience, and Tone

Learning Objective Upon completing this chapter, you will be able to identify your communication purpose and your audience, thereby achieving the appropriate tone in every workplace writing situation.

Every instance of workplace writing occurs for a specific reason and is intended for a particular individual or group. Much the same is true of spoken messages, whether delivered in person or over the telephone. Therefore, there is always both a purpose and an audience to take carefully into account, to ensure that the tone of the exchange will be appropriate to the situation. While this may seem obvious, awareness of purpose, audience, and tone is the single most crucial factor determining whether your communication will succeed. This opening chapter concentrates on these fundamental concerns, presents a brief overview of the basic principles involved, and provides exercises in their application.

Purpose

Nearly all workplace writing is done for one or more of three purposes: to create a record, to request or provide information, or to persuade. A caseworker in a social services agency, for example, might interview an applicant for public assistance to gather information that will then be reviewed in determining the applicant's eligibility. Clearly, this is an instance of writing that is intended both to provide information and to create a record. The purchasing director of a manufacturing company, on the other hand, might write a letter inquiring whether a particular supplier can provide materials more cheaply than the current vendor. Presumably, the inquiry will receive a prompt written reply. Obviously, the primary purpose of both letters is to exchange information. In yet another setting, a probation officer composes a pre-sentencing report intended to influence the court to grant probation to the offender or impose a jail sentence. The officer may recommend either, and the report will become part of the offender's record, but the primary purpose of this example of workplace writing is to persuade.

The first step in the writing process is to consciously identify which of the three categories of purpose applies. You must ask yourself, "Am I writing primarily to create a record, to request or provide information, or to persuade?" Once you make this determination, the question becomes, "Summarized in one sentence, what am I trying to say?" To answer, you must zoom in on your subject matter, focusing on the most important elements. A helpful strategy is to employ the "5 W's" that journalists use to structure the opening sentences of newspaper stories: Who, What, Where, When, Why. Just as they do for reporters, the 5 W's will enable you to get off to a running start. Consider, for example, how the 5 W's technique applies in each of the following situations:

- ***Caseworker writing to provide information and create a record***

  WHO WHAT WHERE
  Carolyn Matthews visited the downtown office of the County
  WHEN WHY
  Social Services Department on May 15 to apply for public assistance.

- ***Purchasing Director writing to request information***

  WHO WHAT
  I'd like to know whether you can provide gaskets for less than
  WHERE WHEN
  $100/dozen, shipped to my company on a monthly basis,
  WHY
  because I am seeking a new supplier.

- ***Probation Officer writing to persuade***

  WHO WHAT
  Jerome Farley should be denied probation and sentenced to
  WHERE WHEN WHY
  state prison, effective immediately, because he is a repeat offender.

Audience

Next ask yourself, "Who will read what I have written?" This is a crucial aspect of the communications process. To illustrate, consider these introductory paragraphs from two recent articles on the subject of breast cancer. The first excerpt is from *Good Housekeeping,* a popular monthly magazine, while the second is from *Cancer,* the medical journal of the American Cancer Society.

> In the last five years or so, we've made a great deal of progress in breast cancer prevention. While we still have more questions than answers, the questions themselves are significant—and we're closer to the answers than we've ever been.

The issues of prevention are growing more important as we're becoming more able to identify people who have hereditary and genetic breast cancer. That isn't theoretical risk, but actual risk. The hope is that someday we'll be able to say, for example, "You have a fifty percent risk by age forty and an eighty percent risk by age eighty." Once someone knows that, she's really going to want to find some form of prevention.

From Susan Love, M.D., "Your Best Self-Defense Against Breast Cancer." *Good Housekeeping* (May 1995), p. 72.

To be useful, breast-conserving therapy must provide not only survival equivalent to mastectomy but also low rates of local recurrence, satisfactory cosmetic results, and a low risk of complications. Prospective randomized trials have established that breast-conserving therapy and mastectomy provide equivalent survival rates. The precise criteria required to assure a low rate of local recurrence, however, are still controversial. Various factors have been identified that affect the rate of recurrence in the breast after breast-conserving therapy. Among these, the major factors are the presence or absence of carcinoma at the inked margins of resection (margins), the volume of excision, and the presence or absence of an extensive intraductal component (EIC).

From Irene Gage, M.D., *et al.*, "Pathologic Margin Involvement and the Risk of Recurrence in Patients Treated with Breast-Conserving Therapy." *Cancer* 78.9 (1 November 1996), pp. 1921–1928.

Anyone can immediately recognize the differences between these two pieces of writing. Obviously, the *Good Housekeeping* coverage is general in nature, employs simple vocabulary and no technical terms, and is therefore easy to follow. The *Cancer* article, on the other hand, with its highly specialized content and terminology, is much more challenging. Even the *titles* of the two articles reflect these contrasts. The reason for the differences is that a popular magazine like *Good Housekeeping* is intended for the general public, while a professional journal like *Cancer* is written specifically for highly educated experts. Both articles were written by recognized authorities (the author of the first article was also a co-author of the second), and the purpose of both articles is to inform. But the two publications are targeted at entirely different audiences. Hence the dramatic dissimilarity between the two excerpts. This dissimilarity makes sense. For the *Good Housekeeping* piece to be significantly more sophisticated, or for the *Cancer* piece to be any less so, would be inappropriate. Each is well suited to its readership.

Workplace communications are governed by this same dynamic. A memo, letter, report, or oral presentation must be tailored to its intended audience; otherwise it probably will not achieve the desired results. Therefore, ask yourself the following questions before attempting to prepare any sort of formal communication:

- Am I writing to one person or more than one?
- What are their job titles and/or areas of responsibility?
- What is the level of their education and/or technical expertise?
- What do they already know about the topic?
- Why do they need this information?
- What do I want them to do as a result of receiving it?
- What factors may influence their response?

Obviously, these questions are closely related, so the answers will sometimes overlap. Nevertheless, by profiling your readers or listeners in this way, you will come to see the subject of your planned communication from the viewpoint of your audience as well as from your own. You will be better able to state its purpose, provide necessary details, cite meaningful examples, achieve the correct level of formality, and avert possible misunderstandings, thereby achieving your desired outcome.

In identifying your audience, remember that workplace communications fall into four broad categories:

- ***Upward communication:*** Intended for those above you in the workplace hierarchy.
- ***Lateral communication:*** Intended for those at your own level.
- ***Downward communication:*** Intended for those below you in the hierarchy.
- ***Outward communication:*** Intended for those outside your workplace.

These differences will influence your communications in many ways, particularly by determining format. For in-house communications (the first three categories) the memo traditionally is the preferred medium, with electronically generated (e-mail) memos rapidly growing in popularity in workplaces that have in-house computer networks. For outward communication, such as correspondence with clients, customers, or the general public, the standard business letter has traditionally been the norm. Business letters are either mailed or transmitted by a fax machine. If, however, you are corresponding with another workplace with which you are linked via computer network, e-mail is obviously the fastest, most efficient choice.

Tone

Your hierarchical relationship to your reader will play a major role in determining the *tone* of your communication as well. This is especially true when you are attempting to convey "bad news" (the denial of a request from an employee whom you supervise, for example) or to suggest that staff members adopt some new or different procedure. Although such messages can be phrased in a firm, straightforward manner, a harsh voice or belligerent attitude is seldom productive.

The workplace is essentially a set of individuals and relationships, busy people working together to accomplish a common goal: the mission of the business, organization, or agency. A high level of cooperation and collective commitment is needed in order for this to happen. Ideally, each person exerts a genuine effort to foster a climate of shared enthusiasm and commitment. When co-workers become defensive or resentful, morale problems inevitably develop, undermining productivity. In such a situation, everyone loses.

Therefore, do not try to sound tough or demanding when writing memos about potentially sensitive issues. Instead, appeal to the reader's sense of fairness and cooperation. Phrase your sentences in a non-threatening way, emphasizing the reader's point of view by using a reader-centered (rather than a writer-centered) perspective. For obvious reasons, this approach should govern your correspondence intended for readers outside the workplace.

Here are some examples of how to create a reader-centered perspective by means of creative revision:

Writer-Centered Perspective	**Reader-Centered Perspective**
If I can answer any questions, I shall be happy to do so.	If you have any questions, please ask.
We shipped the order this morning.	Your order was shipped this morning.
I am happy to report that . . .	You will be glad to know that . . .

Note the use of "your," to personalize the communication, which we'll refer to as the "you" approach. Always remember "please," "thank you," and other polite terms.

Now consider Figures 1.1 and 1.2. Both memos have the same purpose, to change a specific behavior, and both address the same audience.

EASTERN MANUFACTURING, INC.

MEMORANDUM

DATE: March 10, 1999

TO: All Employees

FROM: Brian Johnson, Supervisor
Maintenance

SUBJECT: Littering

Since January 1, smoking is strictly prohibited inside the Main Building. Do NOT smoke anywhere indoors!

Some of you still insist upon smoking, and have been doing so outside. As a result, the areas near the rear exit and around the picnic tables are constantly littered with smoking-related debris (filter tips, half-smoked cigarettes, matchbooks, etc.), creating an eyesore and making more work for my staff, who have to keep cleaning up this mess.

Starting Monday, sand buckets will be provided outside the rear doors and in the picnic area. Use them!

FIGURE 1.1 Original Memo

But the first version adopts a writer-centered approach and is harshly combative. The reader-centered revision, on the other hand, is diplomatic and therefore much more persuasive. The first is almost certain to create resentment and hard feelings, while the second is far more likely to achieve the desired results.

EASTERN MANUFACTURING, INC.

MEMORANDUM

DATE: March 10, 1999

TO: All Employees

FROM: Brian Johnson, Supervisor
Maintenance

SUBJECT: Outdoor Ashtrays

Now that the Main Building has become a No Smoking zone, some of you have been taking your breaks outdoors.

We appreciate your compliance with the new regulations, and wish to minimize your inconvenience. As of Monday, sand bucket "ashtrays" will be provided for your use outside the rear doors and near the picnic tables. This will help maintain a more pleasant atmosphere for us all, by minimizing litter behind the building.

Again, thanks very much for your cooperation!

FIGURE 1.2 Revised Memo

In most settings, you may adopt a somewhat more casual manner with your equals and with those below you than with those above you in the chain of command, or with persons outside the organization. But in any case avoid an excessively conversational style. Even when the situation is not particularly troublesome, and even when your reader is well known to you, remember that "business is business." While you need not sound stuffy, it is important to maintain a certain level of formality. Accordingly, you should never allow personal matters to appear in workplace correspondence. Consider, for example, Figure 1.3, a memo in which the writer has obviously violated this rule. Although the writer's tone toward his supervisor is appropriately respectful, the content of his memo should be far less detailed. The revised version in Figure 1.4 is therefore much better.

A slangy, vernacular style is also out of place in workplace writing, as are expletives and any coarse or vulgar language. Something that may seem clever or humorous to you may not amuse your reader, and will probably appear foolish to anyone reviewing the correspondence later on. Keep this in mind when sending computer-generated messages via e-mail, a medium that seems to encourage a looser, more playful manner of interaction. Typical of this are e-mail emoticons, which are silly "faces" created by combinations of punctuation marks, like these:

: -)	: - (	; -)
Smile	Frown	Wink

Briefly popular when first devised, they now distract or annoy most serious-minded readers, undermining the writer's credibility. Stay away from them.

A sensitive situation awaits you when you must convey unpleasant information or request assistance or cooperation from superiors. Although you may sometimes yearn for a more democratic arrangement, there is within every workplace a pecking order that must be taken into account as you choose your words. Hierarchy exists because some individuals—by virtue of greater experience, education, or access to information—are in fact better positioned to lead. Although this system sometimes functions imperfectly, the supervisor, department head, or other person in charge will respond better to subordinates whose communications reflect an understanding of this basic reality. Essentially, the rules for writing to those higher on the ladder are the same as for writing to those on the lower rungs. Be focused and self-assured, but use

NORTHERN INDUSTRIES

MEMORANDUM

DATE: April 15, 1998

TO: Marilyn Grant, Supervisor
Shipping Dept.

FROM: Paul Curtis
Shipping Dept.

SUBJECT: Personal Leave Request

I have three days of personal leave saved up, and am asking your permission to take off from work next Monday, Tuesday, and Wednesday (April 20–22).

As you probably have heard, I've been having a lot of family problems lately. My son was recently arrested for drug possession, and my wife is talking about leaving. I really need a few days off to try to get my home situation straightened out.

Please approve this request.

FIGURE 1.3 Original Memo

the "you" approach, encouraging the reader to see the advantage in accepting your recommendation or granting your request.

An especially polite tone is advisable when addressing those who outrank you. Acknowledge that the final decision is theirs, and that you

NORTHERN INDUSTRIES

MEMORANDUM

DATE: April 15, 1998

TO: Marilyn Grant, Supervisor
Shipping Dept.

FROM: Paul Curtis
Shipping Dept.

SUBJECT: Personal Leave Request

I have three days of personal leave saved up, and am asking your permission to take off from work next Monday, Tuesday, and Wednesday (April 20–22).

Because I have some urgent personal business to attend to, I would certainly appreciate your approving this request. It's really quite important.

Thank you very much for your consideration.

FIGURE 1.4 Revised Memo

are fully willing to abide by that determination. This can be achieved either through "softening" words and phrases ("perhaps," "with your permission," "if you wish") or simply by stating outright that you will accept whatever outcome may develop. Consider, for example, the memos in Figures 1.5 and 1.6. Although both say essentially the same thing, the first is completely inappropriate in tone, so much so that it would likely result in negative personal consequences for the writer. The second would be much better received because it properly reflects the nature of the professional relationship between writer and reader.

Western Trucking, Inc.

MEMORANDUM

DATE: May 18, 1998

TO: Anne Scott, Supervisor
Dispatching

FROM: Thomas Kearney, Driver

SUBJECT: Drug Testing

Just wanted to let you know that you'd better forget about the random drug-testing policy you announced in your memo yesterday. It's a dumb idea that will never work. All the drivers are angry about it, and there are a lot of unanswered questions that your memo left completely unclear! From what I hear, people in other departments have a lot of questions too. Better clear some of this stuff up or nobody's ever going to hold still for it.

FIGURE 1.5 **Original Memo**

Communicating with customers or clients also requires a great deal of sensitivity and tact. When justifying a price increase, denying a claim, or apologizing for a delay, you will probably create an unpleasant climate unless you present the facts in an unantagonistic manner. Always strive for the most upbeat, reader-centered wording you can devise. Here

Western Trucking, Inc.

MEMORANDUM

DATE: May 18, 1998

TO: Anne Scott, Supervisor
Dispatching

FROM: Thomas Kearney, Driver

SUBJECT: Drug Testing

There is some confusion about the new drug testing policy that was announced yesterday. Probably as a result of that misunderstanding, there also appears to be some resistance to the plan.

If you'll permit me a suggestion, it may be a good idea to schedule a brief meeting with the employees to offer information, address their concerns, and clarify some of the more troubling features of the policy.

Thank you for considering this idea, and please let me know if I can assist in any way.

FIGURE 1.6 **Revised Memo**

are some examples of how to rephrase negative content in more positive, reader-centered terms:

Negative Wording	**Positive Wording**
We cannot process your claim because the necessary forms have not been completed.	Your claim can be processed as soon as you complete the necessary forms.
We do not take phone calls after 3 p.m. on Fridays.	You may reach us by telephone until 3 p.m. on Fridays.
We closed your case because we never received the information requested in our letter of April 2.	Your case will be reactivated as soon as you provide the information requested in our April 2 letter.

When the problem has been caused by an error or oversight on your part, be sure to apologize. However, do not state specifically what the mistake was, or your letter may be used as evidence against you should a lawsuit ensue. Simply acknowledge that a mistake has been made, express regret, explain how the situation will be corrected, and close on a conciliatory note. Consider, for example, the letter in Figure 1.7. The body and conclusion are fine, but the introduction practically invites legal action. Here is a suggested revision of the letter's opening paragraph, phrased in less incriminating terms:

> Thank you for purchasing our product and for taking the time to contact us about it. We apologize for the unsatisfactory condition of your Superior microwave dinner.

Moreover, given the serious nature of the complaint, the Customer Services representative should certainly have made a stronger effort to establish a tone of sincerely apologetic concern. As it stands, this letter seems abrupt and rather impersonal—certainly not what the context requires. (For a much better handling of this kind of situation, see the adjustment letter in Figure 2.10.)

By determining your purpose and carefully analyzing your intended audience you will achieve the correct tone for any communication situation. As we have seen, this is crucial when dealing with potentially resistive readers (especially those above you in the workplace hierarchy) and when rectifying errors for which you are accountable. In all instances, however, a courteous, positive, reader-centered approach gets the best results.

Superior Foods, Inc.

135 Grove St., Atlanta, GA 30300 • (404) 555-1234

October 11, 1999
Mr. Philip Updike
246 Alton St.
Atlanta, GA 30300

Dear Mr. Updike:

We are sorry that you found a piece of glass in your Superior microwave dinner. Please accept our assurances that this is a very unusual incident.

Here are three coupons redeemable at your local grocery market for complimentary Superior dinners of your choice.

We hope you will continue to enjoy our fine products.

Sincerely,

John Roth

John Roth
Customer Services Dept.

Enclosures (3)

FIGURE 1.7 **Letter to Customer**

Exercises

■ EXERCISE 1.1

Revise each of the following three memos to achieve a tone more appropriate to the purpose and audience.

Northwestern Manufacturing

MEMORANDUM

DATE: June 7, 1999

TO: Charles Gilliam, Supervisor
Maintenance Department

FROM: Jane Neal, Mechanic
Maintenance Department

SUBJECT: Vacation

You probably don't remember, but I usually take my vacation in July. This year I'm taking the last two weeks of August instead. If you disapprove this you'll create considerable inconvenience for me and you better believe I'll file a grievance.

EXERCISE 1.1 Continued

COUNTY DEPARTMENT OF SOCIAL SERVICES

MEMO

DATE: March 16, 1999

TO: All Caseworkers

FROM: Cheryl Alston, Case Supervisor

SUBJECT: Goofing Off

A lot of you seem to think that this is a country club, and are spending entirely too much time in the break room! As you well know, you're entitled to one fifteen-minute break in the morning and another in the afternoon. The rest of the time you're supposed to be AT YOUR DESK unless signed out for fieldwork.

EXERCISE 1.1 Continued

County Community College

MEMORANDUM

DATE: May 4, 1999

TO: All Employees

FROM: Charles Rigney, Chief
Security

SUBJECT: Burglarized Vehicles

Recently there's been a rash of burglaries in the faculty/staff parking lot. Items such as tape decks, cellular phones, and even a personal computer have been reported missing from vehicles.

Upon investigation, however, we've learned that several of these vehicles had been left unlocked. Don't be stupid! Always lock your car or else be prepared to get ripped off. My staff can't be everywhere at once, and if you set yourself up to be victimized, it's not our fault.

■ EXERCISE 1.2

Revise each of the following three letters to achieve a tone more appropriate to the purpose and audience.

Bancroft's in the Mall

The Turnpike Mall • Turnpike East • Augusta, Maine 04330

February 17, 1998

Ms. Barbara Wilson
365 Grove St.
Augusta, ME 04330

Dear Ms. Wilson:

Your Bancroft's charge account is $650.55 overdue. We must receive a payment immediately.

If we do not receive a minimum payment of $50 within three days, we will refer your account to a collection agency and your credit rating will be permanently compromised.

Send a payment at once!

Sincerely,

Michael Modoski

Michael Modoski
Credit Department

EXERCISE 1.2 Continued

Southeast Insurance Company

Southeast Industrial Park
Telephone: (904) 555-0123

Tallahassee, FL 32301
FAX: (904) 555-3210

November 4, 1997

Mr. Francis Tedeschi
214 Summit Avenue
Tallahassee, FL 32301

Dear Mr. Tedeschi:

This is to acknowledge receipt of your 11/1/97 claim.

Insured persons entitled to benefits under the Tallahassee Manufacturing Co. plan effective December 1, 1996 are required to execute statements of claims for medical-surgical expense benefits only in the manner specifically mandated in your certificate holder's handbook.

Your claim has been quite improperly executed, as you have neglected to procure the Physician's Statement of Services Rendered. The information contained therein is prerequisite to any consideration of your claim.

Enclosed is the necessary form. See that it's filled out and returned to us without delay, or your claim cannot be processed.

Yours truly,

Ann Jurkiewicz

Ann Jurkiewicz
Claims Adjustor

Enclosure

■ EXERCISE 1.2 Continued

DEPARTMENT OF SOCIAL SERVICES

County Administration Building **Easton, NJ 07300**
(201) 555-0123

November 10, 1997

Easton Savings Bank
36 Bank Street
Easton, NJ 07300

Re: Charles Mangan (Social Security # 000-00-0000)

To Whom It May Concern:

The above individual has applied for Medical Assistance. This Department requires that a thirty-month banking history accompany all such applications. You must send us the necessary information immediately.

Provide a listing of each month's average balance for the period of March 1, 1995 to November 1, 1997, along with verification of all closed or transferred accounts during that period.

This directive is made pursuant to New Jersey State Law, which mandates that all banking organizations must furnish such information to authorized representatives of the Department of Social Services to verify eligibility for any form of Public Assistance.

Sincerely,

Mary Louise Martin

Mary Louise Martin
Caseworker

EXERCISE 1.3

Revise each of the following three memos to eliminate inappropriate tone and/or content.

Southern Industries, Inc.

MEMORANDUM

DATE: April 2, 1998

TO: Sandra Spink, Director
Maintenance

FROM: Martin Avery, Shift Supervisor
Maintenance

SUBJECT: Sprinkler System Screw-Up

I don't blame you for being angry about the equipment malfunction that caused the sprinkler system to activate while the Mayor's Blue Ribbon Commission was touring the plant yesterday.

All I can tell you is, it sure wasn't MY fault! A couple of the technicians were running a routine check on the system and somehow set the stupid thing off by accident. No big deal. But I did tell them that if anything like this ever happens again they'll be dead meat. Maybe we should send an apology to the Mayor so the phony old windbag doesn't badmouth us to the media.

EXERCISE 1.3 Continued

EASTERN MANUFACTURING, INC.

MEMORANDUM

DATE: July 7, 1998

TO: Richard Rhodes, Supervisor
Purchasing

FROM: Jodi Mueller, Secretary
Purchasing

SUBJECT: Excuse for Absence

Sorry I missed work on Monday. What happened was that my husband's company picnic was on Sunday. As you may have heard, he has a really bad drinking problem. Needless to say, he tied one on big-time, and insisted on staying out half the night, so I didn't get any sleep. But I gave him a good talking-to, and I can promise you that nothing like this will ever happen again.

EXERCISE 1.3 Continued

The Centerton Company

MEMORANDUM

DATE: November 18, 1998

TO: All Marketing Dept. Employees

FROM: Carl Roberts, Supervisor
Marketing

SUBJECT: Rescheduling of Meeting

The Friday afternoon department meeting has been rescheduled for Monday at 9 a.m., as I have to leave work early on Friday.

My son's high school football team (the mighty 7 & 0 Centerton Lions—rah! rah!) have an out-of-state game Friday night against another undefeated team, in Illinois. From what I understand, they're a real powerhouse, but I'm sure Centerton will beat them, especially since Carl Junior's averaging nearly fourteen yards per carry!

: -)   : -)   : -)   : -)   : -)   : -)   : -)

GO, LIONS!!!

■ EXERCISE 1.4

Revise each of the following three letters to eliminate wording that might create legal liability.

133 Court Street Olympia, WA 98501

January 11, 1999

Mr. Robert Ryan
352 Stegman Street
Olympia, WA 98501

Dear Mr. Ryan:

We have received your letter of January 3, and we regret that the heating unit we sold you has malfunctioned, killing $1,500 worth of your tropical fish.

Since the unit was purchased more than three years ago, however, our store-wide warranty is no longer in effect and we are therefore unable to accept any responsibility for your loss. Nevertheless, we are enclosing a Fin & Feather discount coupon good for $10 toward the purchase of a replacement unit or another product of your choice.

We look forward to serving you in the future!

Sincerely,

Sandra Kouvel

Sandra Kouvel
Store Manager

Enclosure

■ EXERCISE 1.4 Continued

TELEVISION WORLD

521 Scott Street Ames, Iowa 50010 (515) 555-1234

February 20, 1999

Ms. Christine Nguyen
230 Fairview Street
Ames, Iowa 50010

Dear Ms. Nguyen:

Thank you for your recent letter about the faulty wiring in the television set you purchased at Television World. We are glad to hear that the fire it caused resulted in only minor damages to your apartment.

If you will bring the television in we'll gladly exchange it for a more reliable set. Customer satisfaction is our #1 priority!

We are happy to assist you with all your video needs.

Yours truly,

Peter Keane

Peter Keane
Store Manager

■ EXERCISE 1.4 Continued

HIGH ROLLER
BIKES & BOARDS

516 Bridge Street ■ Phoenix, AZ 85000

August 17, 1999

Mr. Scott Damsky
252 Sheridan Street
Phoenix, AZ 85000

Dear Mr. Damsky:

We are sorry that the bicycle tire we sold you burst during normal use, causing personal injury resulting in lingering lower back pain.

Certainly we will install a replacement tire free of charge if you simply bring your bicycle into our shop any weekday during the hours of 9 a.m. to 5 p.m.

Thank you for purchasing your bicycle supplies at High Roller!

Sincerely,

Robin Coolidge

Robin Coolidge
Store Manager

2

Correspondence: Memos and Letters

Learning Objective

Upon completing this chapter, you will be able to use basic format and organization patterns to write effective memos and letters.

Of all the forms of written communication used in the workplace, memos (including e-mail) and business letters are by far the most common. Any large corporation, agency, or other organization generates thousands of such documents daily. And even in a small setting, they are fundamental to office procedure. Focusing on both format and content, and exploring some of the effects of recent technological advances, this chapter explains how to handle these routine but essential forms of correspondence.

Memos

Traditionally, the memo has been a vehicle for internal or "intramural" communication—a message from someone at Company X to someone else at Company X. The memo may be written to one person or to a group, but until fairly recently it has almost always been a form of in-house correspondence.

In such a situation the writer and reader(s) may be well acquainted. They may have seen one another moments before or even had lunch together. Indeed, the contents of the memo may already be known to all parties involved in the exchange. Although the usual purpose of a memo is to inform, often its function is to create a written record of a request or other message previously communicated in person, over the telephone, or through the grapevine.

Accordingly, a memo is usually quite direct in approach. While an introductory sentence or two may be helpful to orient the reader, you should come to the point quickly and not ramble on. A common error is to obscure the central issue by confusing the reader with irrelevant details. A good memo focuses sharply, zooming in on what the reader needs to know. Depending on the subject, you should certainly be able to accomplish this in three or four short paragraphs, and one is often enough.

Format

There is essentially one basic format for a memo. Although minor variations do indeed exist, practically all memos—including e-mail generated on the computer—share certain standard features in formatting:

- The word "Memo," "Memorandum," or some equivalent term, at or near the top of the page.

- The TO line, enabling the memo to be "addressed," and FROM line, allowing it to be "signed." Always use the full name, title, and/or department of the person to whom you are writing. This not only ensures that the memo will reach its intended destination, but also creates a more complete record for anyone reviewing the file later. For the same reason, use your own full name, title, and/or department. Be aware also that the TO and FROM lines eliminate the need for a salutation ("Dear Ms. Bernstein") and a complimentary close ("Yours truly"); although some writers like to use these devices as a way of making their memos seem less impersonal, they are by no means necessary and in most instances are better omitted.
- The DATE line (provided automatically on e-mail, along with the exact *time* of transmission).
- The SUBJECT line, identifying the topic. Like a newspaper headline, but even more concisely, the SUBJECT line orients and prepares the reader for what is to follow. To write a good subject line, answer this question, In no more than three words, what is this memo really about?
- And, of course, the message or content of the memo. As explained earlier, three or four paragraphs should be sufficient: a concise introduction, a middle paragraph or two conveying the details, and perhaps a brief conclusion. If the message is quite simple, however, you should get to the point quickly. Some memos are as short as one paragraph, or even one sentence. Like so many other features of workplace communication, memo length is determined by purpose and audience.

The memo in Figure 2.1 embodies all of these features, and also provides an opportunity to explore further the principle of *tone* introduced in Chapter 1.

The personnel manager has picked her words carefully, to avoid sounding bossy. She says "You *may want* to send him a . . . card," not "You *should* send him a . . . card," even though that's what she really means. As discussed in Chapter 1, a tactful writer can soften a recommendation, a request, or even a command simply by phrasing it in a diplomatic way. In this situation an employee's decision whether to send a card is strictly a matter of personal choice, so the memo's gentle tone is particularly appropriate. But the same strategy can also be used when conveying important directives you definitely expect the reader to follow.

CITY MANUFACTURING CO.

MEMORANDUM

DATE: May 15, 2000

TO: All Employees

FROM: Sarah Williams, Manager
Personnel Department

SUBJECT: Greg Massey

As many of you already know, Greg Massey of the Maintenance Department was admitted to Memorial Hospital over the weekend and is scheduled to undergo surgery on Tuesday.

Although Greg will not be receiving visitors or phone calls for a while, you may want to send him a "Get Well" card to boost his spirits. He's in Room 325.

We'll keep you posted about Greg's progress.

FIGURE 2.1 **Basic Memo Format**

E-Mail

During the past decade, most employers have acquired computerized communications systems that enable workers to use e-mail (memos written and transmitted electronically, and viewed on the computer screen rather than on paper). Although many different e-mail systems are available, all operate in much the same fashion. Typically, the writer logs in, typing his or her name, along with a confidential password that prevents unauthorized access. After the writer types a short sequence of *commands,* a memo form called a template appears on the screen. The writer simply fills in the blanks, as if typing on a pre-printed memo form. When the message is completed, with the touch of a key it can then immediately be sent to as many other users of the system as the writer wishes—one or everyone. The memo is also stored in the reader's own electronic mailbox, and can be kept there indefinitely for future reference. To read incoming memos, a similar procedure is used. After logging on, the reader types a slightly different series of responses, whereupon each memo received can be viewed on the screen. Depending on the reader's preferences, each memo then can either be deleted, filed for future reference, printed, answered, forwarded, or a combination of these options. Figure 2.2 is an e-mail memo.

There are important advantages to electronic information technology. On the most obvious level, e-mail is incomparably faster than conventional correspondence. In the past, at least five distinct steps were involved in communicating by memo:

1. writing;
2. typing by secretary;
3. proofreading by writer;
4. photocopying for writer's file;
5. routing to intended reader(s).

Depending on office workload and clerical staffing levels, this could be a very time-consuming process. With e-mail, however, all five steps are compressed into one, permitting rapid communication.

E-mail is still in its relative infancy, however, and will not reach its full potential until certain intrinsic problems have been resolved. One major drawback is that the very ease with which e-mail can be generated encourages overuse. In the past, a writer would not bother to send a memo without good reason; there was too much time and effort involved to do otherwise. Now, however, much needless correspondence is produced. Yesterday's writers would wait until complete information on a given topic had been received and organized before passing it along

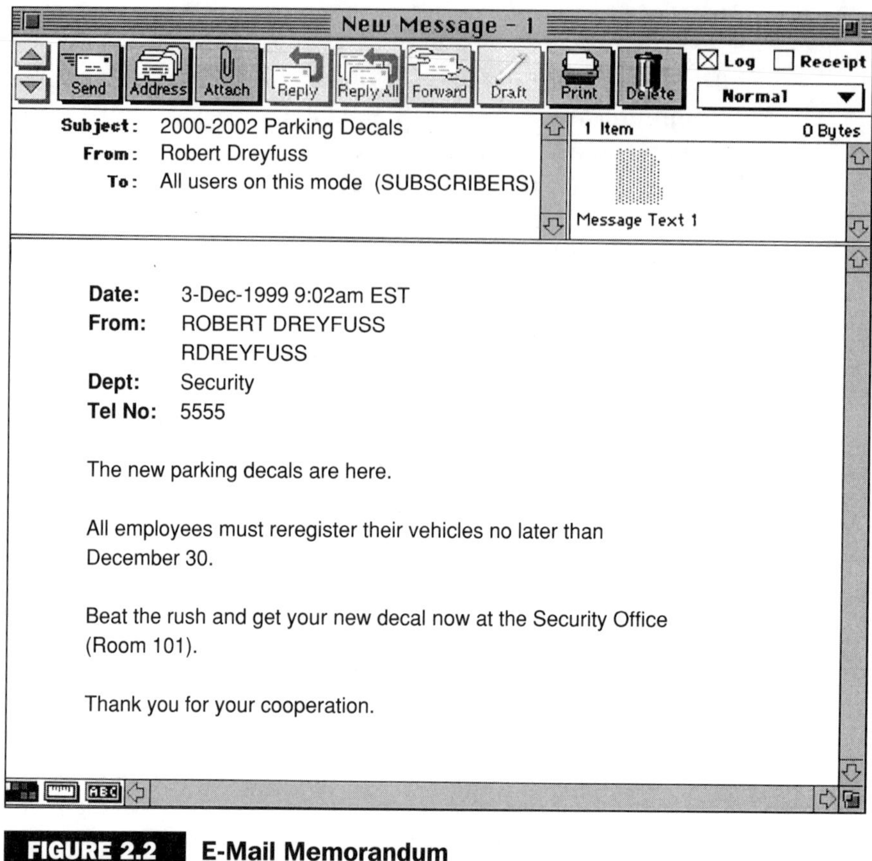
New Message - 1

Send | Address | Attach | Reply | Reply All | Forward | Draft | Print | Delete

Log | Receipt | Normal

Subject: 2000-2002 Parking Decals
From: Robert Dreyfuss
To: All users on this mode (SUBSCRIBERS)

1 Item 0 Bytes
Message Text 1

Date: 3-Dec-1999 9:02am EST
From: ROBERT DREYFUSS
RDREYFUSS
Dept: Security
Tel No: 5555

The new parking decals are here.

All employees must reregister their vehicles no later than December 30.

Beat the rush and get your new decal now at the Security Office (Room 101).

Thank you for your cooperation.

FIGURE 2.2 E-Mail Memorandum

to others. But today it is not uncommon for many memos to be written on the same subject, doling out the information in bits and pieces, sometimes within a very short time span. The resulting fragmentation wastes the energies of writer and reader alike and increases the possibility of confusion, not to mention the likelihood of premature response.

Similarly, memos about sensitive issues are often dashed off in "the heat of battle" without sufficient reflection. In the past, the writer always had some time to reconsider a situation before actually sending a memo, and had the option of revising or simply destroying the memo if, at the proofreading stage, it had come to seem a bit too insistent or otherwise inappropriate. The inherently rapid-fire nature of e-mail, however, all but eliminates any such opportunity for second thoughts. In addition, hasty composition causes a great many keyboarding miscues, omissions, and other fundamental blunders that must then be corrected in subsequent memos, creating the inefficient phenomenon of "e-mail about e-mail." Indeed, the absence of a secretarial "filter" has given rise to a great deal of embarrassingly bad writing in the workplace. You risk ridicule and loss of credibility unless you closely proofread every e-mail message before sending it. Make sure that the information is correct and that all pertinent details have been included. Be particularly careful to avoid typos, misspellings, faulty capitalization, sloppy punctuation, and basic grammatical errors.

As mentioned in Chapter 1, the company e-mail network is no place for personal messages or an excessively conversational style. Many employers provide a separate e-mail "bulletin board" on which workers can post and access announcements about garage or vehicle sales, carpooling, theater tickets, and the like. Such matters are appropriate only as bulletin board content. As use of e-mail has spread, an informal code of etiquette has developed, which most users attempt to observe. Here are three major taboos:

- ***Flaming.*** Refrain from openly hostile or abusive comments, whether directed at the reader or toward a third party. The fact that you are communicating electronically rather than face-to-face does not permit you to violate the basic principles of workplace courtesy.
- ***Screaming.*** Do not compose an entire memo in capital letters.
- ***Spamming.*** Do not send a memo to everyone in the network unless it really applies to everyone. This is especially important when answering a memo that has been sent to everyone on a given mailing list. Unless you have good reason to reply to everyone on that list, *reply only to the sender.*

Now that so many organizations are linked via computer networks, the memo is no longer just an intramural communications medium. Since the memo format is the most common format used "on line," it is beginning to rival the business letter as the major form of correspondence across company boundaries. Clearly, tone takes on even greater importance in e-mail memos sent to readers at other locations. Since the writer and the reader are usually not personally known to each other, a higher level of courteous formality is therefore in order. In addition, the subject matter is often more involved than that of in-house correspondence, so memos sent outside one's workplace are commonly longer and more fully developed than those intended for co-workers. In all other aspects, however, they are essentially the same. Whether in-house or not, whether electronically generated or not, all memos are subject to the principles outlined in the following checklist.

☑ Checklist Evaluating a Memo

A good memo

___ follows the standard format;

___ includes certain features:

- ☐ MEMO or MEMORANDUM as a heading or near the top,
- ☐ TO line, which provides the full name, title, and/or department of the receiver,
- ☐ FROM line, which provides the full name, title, and/or department of the sender,
- ☐ DATE line,
- ☐ SUBJECT line, which provides a clear, accurate, but brief indication of what the memo is about;

___ is organized into paragraphs (one is often enough) that cover the subject fully in a well-organized way;

___ includes no inappropriate content;

___ uses clear, simple language;

___ maintains an appropriate tone, neither too formal nor too conversational;

___ contains no typos or mechanical errors in spelling, capitalization, punctuation, and grammar.

Exercises

■ EXERCISE 2.1

You are the assistant to the personnel manager of a metals fabrication plant. Monday is Labor Day, and most of the 300 employees will be given a paid holiday. The company is under pressure, however, to meet a deadline. Therefore, a skeleton force of forty—all in the production department—will be needed to work the holiday. Those who volunteer will have the option of being paid overtime at the standard time-and-a-half rate or receiving two vacation days. If fewer than forty volunteer, others will be assigned to work on the basis of seniority, with the most recently hired employees chosen first. The personnel manager has asked you to alert affected employees. Write a memo.

■ EXERCISE 2.2

You are a secretary at a regional office of a state agency. Normal working hours for Civil Service employees in your state are 8:30 a.m. to 4:30 p.m., with a lunch break from 12 to 12:30 p.m. During the summer, however, the hours are 8:30 a.m. to 4:00 p.m., with lunch unchanged. Summer hours are in effect from July 1 to September 2. It is now mid-June, and the busy office supervisor has asked you to remind employees of the summer schedule. Write a memo.

■ EXERCISE 2.3

You work in the lumber yard of a building supplies company. Every year on the July 4 weekend, the town sponsors the Liberty Run, a 10K (6.2 mile) road-race. This year, for the first time, local businesses have been invited to enter five-member teams to compete for the Corporate Cup. The team with the best combined time takes the trophy. There will be no prize money involved, but much good publicity for the winners. Since you recently ran the Boston Marathon, the company president wants you to recruit and organize a team. It is now April 20. Better get started. Write a memo.

■ EXERCISE 2.4

You are an office worker at a large paper products company that has just installed an upgraded computer system. Many employees are having difficulty with the new software. The manufacturer's representatives will be on-site all next week to provide training. Since you are studying computer technology, you have been asked to serve as liaison. You must inform your co-workers about the training, which will be delivered in Conference Room 3 from Monday through Thursday in eight half-day sessions (9 a.m. to 12 p.m. and 1 to 4 p.m.), organized alphabetically by workers' last names, as follows: A–B, C–E, F–I, J–M, N–P, Q–SL, SM–T, and U–Z. Workers unable to attend must sign up for one of two make-up sessions that will be held on Friday. You must ensure that everyone understands all of these requirements. Write a memo.

■ EXERCISE 2.5

You are the manager of the employee cafeteria at a printing company. For many years the cafeteria has provided excellent service, offering breakfast from 7 to 8:30 a.m. and lunch from 11 a.m. to 2 p.m. It also serves as a break room, selling coffee, soft drinks, and snacks all day. But the cafeteria is badly in need of modernization. Work is scheduled to begin next Wednesday. Naturally, the cafeteria will have to be closed while renovations are under way. Employees will still be able to have lunch and breaks, however, because temporary facilities are being set up in Room 101 of Building B, a now-vacant area formerly used for storage. The temporary cafeteria will provide all the usual services except for breakfast. Obviously, employees need to know about the situation. Write a memo.

■ EXERCISE 2.6

You are the security chief at a manufacturing company that makes small metal hand tools. The plant employs roughly 100 people. Management has told you that a large number of tools are disappearing. According to company records, the plant produces approximately 50,000 per day, but far fewer are actually being shipped out. After double-checking the figures to ensure their accuracy, you have concluded that pilferage is the only possible explanation. A metal detector positioned at the employee exit near the time clock would catch anyone trying to smuggle tools out of the factory. Since the purchase cost of a metal detector is prohibitive, you have decided to rent one. Anyone caught stealing will immediately be fired, and a note to that effect will become part of the individual's personnel file. You don't want to create an atmosphere of hostility, but you do need to inform the employees about these developments. Write a memo.

■ EXERCISE 2.7

You are a caseworker at a new county agency that assists troubled youths by placing them in group homes run by the agency. There are five boys or girls per home, supervised by specially trained counselors. You find this job rewarding, although it involves more paperwork than you'd prefer. Yesterday, for example, the agency psychiatrist recommended a medication change for a boy who resides at Group Home #6. The boy has been diagnosed as hyperactive, and has been receiving a daily dosage of 30 mg of Ritalin (one 10 mg tablet in the morning, one at noon, and one at bedtime). The doctor has decided to increase the dosage to 35 mg daily, by adding a 5 mg morning tablet. You have no reason to question the doctor's judgment, but you must inform the boy's counselors. Write a memo.

■ EXERCISE 2.8

You are the United Way representative at your place of employment, and must therefore encourage everyone to contribute. While no one can be required to donate, your own supervisor has publicly stated that the company goal is 100 percent participation, and you know from past experience that she is intolerant of failure. It is in your own interest, then, to persuade all your co-workers to contribute, but without appearing to pressure them. Write a memo.

■ EXERCISE 2.9

You are the production manager for a computer parts manufacturer. Last month four machines had excessive downtime. The company's production of Part #Z43 has dropped. Two of your best customers have complained about late shipments of Part #Z43. One customer has canceled a standing order and is now buying the part from your principal competitor. For the past two months the company's production of Part #Y01 has also been declining. To discuss the situation, all production supervisors will meet in Conference Room G, in the west wing of the main building, at 10 a.m. next Monday. Each should bring to the meeting up-to-date figures on costs, equipment, personnel, etc. You must inform the production supervisors about the meeting. Write a memo.

EXERCISE 2.10

Proofread and rewrite the following memo, correcting all typos and mechanical errors.

Memorial Hospital

M E M O R A N D U M

DATE: September 16, 1999

TO: All Employes

FROM: Roger Sammon, Clerk
Medical Recrods Department

SUBJECT: Ethel Townsend

As many of you allready know. Ethel Townsand from the Medical records Depratment is retiring next month. After more then thirty years of faithfull service to Memorial hospital.

A party is being planed in her honor. It will be at seven oclock on friday October 30 at big Joes Resturant tickets are $30 per person whitch includes a buffay diner and a donation toward a gift.

If you plan to atend please let me no by the end of this week try to get you're check to me by Oct 20

Letters

Unlike memos, business letters are typically used for *external* communication, a message from someone at Company X to someone elsewhere—a customer or client, perhaps, or a counterpart at Company Y. But there are some similarities between the memo and the letter: the writer and the reader may or may not be acquainted, the message may or may not be news to the reader, and sometimes the objective is simply to create a written record. Usually, however, there is a more immediate purpose. Literally millions of letters are written every day, for an enormous variety of reasons. Some of the more typical purposes of a letter are to:

- ask for information (inquiry);
- sell a product or service (sales);
- purchase a product or service (order);
- request payment (collection);
- voice a complaint (claim);
- respond to a complaint (adjustment);
- thank someone (acknowledgment).

Figures 2.4–2.13 provide examples of all of these purposes. Notice, however, that regardless of purpose, each letter has been formatted in accordance with one of the three most common styles: the modified block style with indented paragraphs, the modified block style, and the full block style.

Modified Block Style With Indented Paragraphs

As shown in Figures 2.4, 2.5, and 2.9, the date line, the complimentary close, and the writer's identification all begin at the center of the page. The first line of each paragraph is indented five spaces. All other lines are flush with the left-hand margin.

Modified Block Style

As shown in Figures 2.6–2.8, the date line, the complimentary close, and the writer's identification all begin at the center of the page. All other lines (including the first line of each paragraph) are flush with the left-hand margin.

Full Block Style

As shown in Figures 2.10–2.13, every line (including the first line of each paragraph) is flush with the left-hand margin.

The three styles share several features: all are single-spaced throughout (except between the separate elements, where double-spacing is used), are centered on the page, and are framed by margins of 1 to 1½ inch. All three styles are in common use, with the modified block style with indented paragraphs considered the most traditional format (and a bit old-fashioned). Full block format, on the other hand, is the most contemporary. As the template included with popular word processing software, such as WordPerfect and Microsoft Word, it is rapidly becoming the norm. Regardless of format, however, every letter includes certain essential components that are set forth on the page in the following sequence:

1. Writer's address (sometimes pre-printed on letterhead) at the top of the page;
2. Date (like e-mail memos, letters sent by fax are automatically imprinted with the exact *time* of transmission as well);
3. Inside address (the full name, title, and address of the receiver);
4. Salutation, followed by a colon (avoid gender-biased salutations such as "Dear Sir" or "Gentlemen");
5. Body of the letter, using the three-part approach outlined below;
6. Complimentary close ("Sincerely" is best), followed by a comma;
7. Writer's signature;
8. Writer's name and title, beneath the signature;
9. Enclosure line, if necessary, to indicate item(s) accompanying the letter.

Along with these standard components, all business letters—irrespective of format—also embrace the same three-part pattern of organization:

1. A brief introductory paragraph establishing context (by referring to previous correspondence, perhaps, or by orienting the reader in some other way) and stating the letter's purpose concisely;
2. A middle section (as many paragraphs as needed) conveying the content of the message by providing all necessary details, presented in the most logical sequence;
3. A brief concluding paragraph politely requesting action, thanking the reader, or providing any additional information pertinent to the situation.

A fairly recent development is the "open punctuation" system in which the colon after the salutation and the comma after the complimentary close are omitted. Figure 2.12 illustrates this variation, which is gaining widespread acceptance. A more radical change is the trend toward a fully abbreviated, "no punctuation/all capitals" approach to the inside address (see Figure 2.13). This derives from the United States Postal Service recommendation that envelopes be so addressed to facilitate computerized scanning and sorting. As the inside address has traditionally matched the address on the envelope, such a feature may well become standard, at least for letters sent by mail rather than by electronic means. Indeed, many companies using "window" envelopes have already adopted this style. For a list of standard abbreviations used in letter writing and formatting, see Figure 2.3.

FIGURE 2.3 Standard Abbreviations

Alabama	AL	Kentucky	KY	Ohio	OH
Alaska	AK	Louisiana	LA	Oklahoma	OK
Arizona	AZ	Maine	ME	Oregon	OR
Arkansas	AR	Maryland	MD	Pennsylvania	PA
California	CA	Massachusetts	MA	Puerto Rico	PR
Colorado	CO	Michigan	MI	Rhode Island	RI
Connecticut	CT	Minnesota	MN	South Carolina	SC
Delaware	DE	Mississippi	MS	South Dakota	SD
District of Columbia	DC	Missouri	MO	Tennessee	TN
		Montana	MT	Texas	TX
Florida	FL	Nebraska	NE	Utah	UT
Georgia	GA	Nevada	NV	Vermont	VT
Hawaii	HI	New Hampshire	NH	Virginia	VA
Idaho	ID	New Jersey	NJ	Washington	WA
Illinois	IL	New Mexico	NM	West Virginia	WV
Indiana	IN	New York	NY	Wisconsin	WI
Iowa	IA	North Carolina	NC	Wyoming	WY
Kansas	KS	North Dakota	ND		
Avenue	AVE	Expressway	EXPY	Parkway	PKWY
Boulevard	BLVD	Freeway	FWY	Road	RD
Circle	CIR	Highway	HWY	Square	SQ
Court	CT	Lane	LN	Street	ST
Turnpike	TPKE				
North	N	West	W	Southwest	SW
East	E	Northeast	NE	Northwest	NW
South	S	Southeast	SE		
Room	RM	Suite	STE	Apartment	APT

Source: United States Postal Service.

The Weekly News

P.O. Box 123
Littleton, New York 13300
Telephone (315) 555-1234 • Fax (315) 555-4321

February 23, 1999

Chief Bernard Rodgers
Littleton Police Department
911 Main Street
Littleton, NY 13300

Dear Chief Rodgers:

It is our understanding that a Littleton resident, Mr. Charles Cartwright, is the subject of an investigation by your department, with the assistance of the County District Attorney. In keeping with the provisions of the New York Freedom of Information Law, I am requesting information about Mr. Cartwright's arrest.

This information is needed to provide our readership with accurate news coverage of the events leading to Mr. Cartwright's current situation. The Weekly News prides itself on fair, accurate, and objective reporting, and we are counting on your assistance as we seek to uphold that tradition.

Since the police blotter is by law a matter of public record, we will appreciate your full cooperation.

Sincerely,

Jane Smith

Jane Smith, Reporter

FIGURE 2.4 Inquiry Letter in Modified Block Style With Indented Paragraphs

254 Sunset Blvd, Weston, CA 95800 • telephone (916) 555-1234

March 3, 1999

Ms. Sarah Levy
643 Glenwood Avenue
Weston, CA 95800

Dear Ms. Levy:

As a preferred customer and holder of our special "Gold Card," you won't want to miss our annual Savings Spectacular.

All the fine clothing pictured in the enclosed brochure has been marked down a full 25%! To take advantage of these incredible bargains, you need only complete the order form on the back cover of the brochure. Or if you prefer, you may simply telephone your order. Our operators are standing by.

Purchases totaling $300 or more are entitled to another 10% off! But you must act quickly! The sale—open to Gold Card customers exclusively—ends on March 10. Order now!

Sincerely,

Jorgé Figueroa

Jorgé Figueroa, Manager
Customer Services Department

Enclosure

FIGURE 2.5 **Sales Letter in Modified Block Style With Indented Paragraphs**

Southton High School

62 Academy Street, Southton, GA 30300
Telephone (404) 555-1234 • Fax (404) 555-4321

July 10, 1999

Value-Rite Office Supplies
462 Decatur Street
Atlanta, GA 30300

Dear Value-Rite:

It's time once again for Southton High to order a shipment of custom-printed, spiral-bound notebooks for use by our students.

You may charge the following order to our account (#2468).

Catalog #	Quantity	Description	Unit Cost	Total
471	300	100 pages	$1.00	$300
472	200	250 pages	2.00	400
473	100	350 pages	3.00	300
			Subtotal	$1000
			Tax (5%)	50
			Shipping	30
			Total	$1080

Please provide blue covers with the gold SHS logo (which you have on file) and ship as promptly as possible.

Sincerely,

Karl Bradbury

Karl Bradbury, Vice-Principal

FIGURE 2.6 Order Letter in Modified Block Style

Greene's

New Acres Mall   Tallahassee, FL 32301

June 15, 1999

Mr. Ernest Cowley
55-A Jackson Road
Tallahassee, FL 32301

Dear Mr. Cowley:

We appreciate your continued patronage of Greene's. We note, however, that your charge account is now $565.31 overdue, and that we have not received your monthly payment since April.

If you have recently sent in your payment, please ignore this friendly reminder. If not, we would appreciate a minimum remittance of $50.00 at your earliest convenience.

If you have any questions about your account, please call us at 555-0123, Ext. 123.

Sincerely,

Heather Sutcliffe

Heather Sutcliffe
Credit Services Department

FIGURE 2.7 **Collection Letter in Modified Block Style**

Jane's Homestyle Restaurants, Inc.

239 Northrop Square Seattle, WA 98100 (206) 555-1234

October 28, 1999

Mr. Joseph Chen, Director
Sales & Service Department
ACE Technologies Corporation
1168 Crosstown Turnpike
Seattle, WA 98100

Dear Mr. Chen:

I purchased the ACE Cash Register System 2000 for my three restaurants in December 1998 and have experienced continuous problems with the video monitors since then.

As recently as September of this year another of the monitors had to be sent in for repairs. Yesterday afternoon that same unit failed again. This occurrence is not uncommon, as you can see by the nine repair invoices I have enclosed for your reference.

Given the many problems we have had with these monitors, I am requesting that you replace them, free of charge. Please call me about this as soon as possible.

Sincerely,

Jane Pelham

Jane Pelham, Owner

Enclosures

FIGURE 2.8 Corporate Claim Letter in Modified Block Style

41 Allan Court
Tucson, AZ 86700

June 29, 1999

Consumer Relations Department
Superior Foods, Inc.
135 Grove Street
Atlanta, GA 30300

Dear Superior Foods:

Superior microwave dinners are excellent products that I have purchased regularly for a number of years. Recently, however, I had an unsettling experience with one of these meals.

While enjoying a serving of Pasta Alfredo, I discovered in the food what appears to be a thick splinter of wood. I'm sure this is an isolated incident, but I thought your Quality Control department would want to know about it.

I've enclosed the splinter, taped to the product wrapper, along with the sales receipt for the dinner. May I please be reimbursed $4.98 for the cost?

Sincerely,

George Eaglefeather

George Eaglefeather

Enclosures

FIGURE 2.9 **Consumer Claim Letter in Modified Block Style With Indented Paragraphs**

Superior Foods, Inc.

135 Grove St., Atlanta, GA 30300 • (404) 555-1234

July 5, 1999

Mr. George Eaglefeather
41 Allan Court
Tucson, AZ 68700

Dear Mr. Eaglefeather:

Thank you for purchasing our product and for taking the time to contact us about it. We apologize for the unsatisfactory condition of your Pasta Alfredo dinner.

Quality is of paramount importance to all of us at Superior Foods, and great care is taken in the preparation and packaging of all our products. Our Quality Assurance staff has been notified of the problem you reported. Although Superior Foods does not issue cash refunds, we have enclosed three complimentary coupons redeemable at your grocery for complimentary Superior dinners of your choice.

We appreciate this opportunity to be of service, and we hope you will continue to enjoy our products.

Sincerely,

John Roth

John Roth
Customer Services Department

Enclosures (3)

FIGURE 2.10 Adjustment Letter in Full Block Style

VALUE-RITE OFFICE SUPPLIES

462 Decatur Street ▪ Atlanta, GA 30300 ▪ (404) 555-1234

March 19, 1999

Ms. Helen Reynard, Owner
Reynard's Auto Palace
Central Highway
Atlanta, GA 30300

Dear Ms. Reynard:

For the past ten years, Value-Rite Office Supplies has purchased all our delivery vans from your dealership, and we have relied upon your service department for routine maintenance and necessary repairs. During that time I have been repeatedly impressed by the professionalism of your employees, especially Jarel Carter, who staffs the service desk.

Both in person and on the telephone, Jarel has always been exceptionally knowledgeable, helpful, and courteous, and is always willing to go the extra mile to ensure customer satisfaction. Just last week, for example, he interrupted his lunch break to get me some information about a part that has been on back-order.

If you can continue to attract employees of Jarel's caliber, you shouldn't have any difficulty remaining the area's #1 dealership. Be sure to keep him in mind the next time you're considering merit raises!

Sincerely,

Richard Cameron

Richard Cameron, Owner

FIGURE 2.11 Acknowledgment Letter in Full Block Style

VALUE-RITE OFFICE SUPPLIES

462 Decatur Street ▪ Atlanta, GA 30300 ▪ (404) 555-1234

March 19, 1999

Ms. Helen Reynard, Owner
Reynard's Auto Palace
Central Highway
Atlanta, GA 30300

Dear Ms. Reynard

For the past ten years, Value-Rite Office Supplies has purchased all our delivery vans from your dealership, and we have relied upon your service department for routine maintenance and necessary repairs. During that time I have been repeatedly impressed by the professionalism of your employees, especially Jarel Carter, who staffs the service desk.

Both in person and on the telephone, Jarel has always been exceptionally knowledgeable, helpful, and courteous, and is always willing to go the extra mile to ensure customer satisfaction. Just last week, for example, he interrupted his lunch break to get me some information about a part that has been on back-order.

If you can continue to attract employees of Jarel's caliber, you shouldn't have any difficulty remaining the area's #1 dealership. Be sure to keep him in mind the next time you're considering merit raises!

Sincerely

Richard Cameron

Richard Cameron, Owner

FIGURE 2.12 **Acknowledgment Letter in Full Block Style With Open Punctuation**

VALUE-RITE OFFICE SUPPLIES

462 Decatur Street ▪ Atlanta, GA 30300 ▪ (404) 555-1234

March 19, 1999

MS HELEN REYNARD
REYNARDS AUTO PALACE
CENTRAL HIGHWAY
ATLANTA GA 30300

Dear Ms. Reynard:

For the past ten years, Value-Rite Office Supplies has purchased all our delivery vans from your dealership, and we have relied upon your service department for routine maintenance and necessary repairs. During that time I have been repeatedly impressed by the professionalism of your employees, especially Jarel Carter, who staffs the service desk.

Both in person and on the telephone, Jarel has always been exceptionally knowledgeable, helpful, and courteous, and is always willing to go the extra mile to ensure customer satisfaction. Just last week, for example, he interrupted his lunch break to get me some information about a part that has been on back-order.

If you can continue to attract employees of Jarel's caliber, you shouldn't have any difficulty remaining the area's #1 dealership. Be sure to keep him in mind the next time you're considering merit raises!

Sincerely,

Richard Cameron

Richard Cameron, Owner

FIGURE 2.13 **Acknowledgment Letter in Full Block Style With Capitalized Inside Address**

As mentioned earlier, more and more companies are communicating with each other via electronic memo rather than by business letter, or—when letters are still used—by fax. Since letters transmitted by fax must be accompanied by a cover memo, it is often more practical to fax the message in memo format. This presupposes, however, that the cover memo and the letter have originated from the same source, and that both writer and reader have ready access to a fax machine. Such is not always the case, especially in situations involving individual clients or customers who still rely on conventional mail delivery. At least for the immediate future, therefore, the letter will remain a major form of workplace correspondence, although its role will continue to undergo redefinition as various forms of electronic communication become increasingly widespread.

Like all successful communication, a good letter must convey an appropriate tone. Obviously, a letter is a more formal kind of communication than a memo because it is more public. Accordingly, a letter should uphold the image of the sender's company or organization by reflecting a high degree of professionalism. But although a letter's style should be somewhat more polished than that of an in-house memo, the language should be no less natural and easy to understand. The key to achieving a readable style—whether in a letter or in anything else you write—is to understand that writing should not sound pompous or "official." Rather, it should sound much like ordinary speech—shined up just a bit. Whatever you do, avoid stilted, old-fashioned business clichés. Strive instead for direct, conversational phrasing. Here is a list of common overly bureaucratic constructions, paired with "plain English" alternatives.

Cliché	**Alternative**
as per your request	as you requested
attached please find	here is
at this point in time	now
in lieu of	instead of
in the event that	if
please be advised that X	X
pursuant to our agreement	as we agreed
until such time as	until
we are in receipt of	we have received
we regret to advise you that X	regrettably, X

If you have a clear understanding of your letter's purpose and have analyzed your audience, you should experience little difficulty achieving the appropriate tone for the situation. If, in addition, your letter has been formatted in accordance with one of the standard layout styles and is written in clear, accessible, and mechanically correct language, the correspondence will likely accomplish its objectives. As noted earlier, you must scrupulously avoid typos and mechanical errors in memos. This is even more important when you compose letters intended for outside readers, who will take their business elsewhere if they perceive you as careless or incompetent. Always proofread carefully, making every effort to ensure that your work is error-free, and consult the following checklist.

☑ Checklist Evaluating a Letter

A good letter

___ follows a standard format;

___ includes certain features:

- ☐ sender's complete address,
- ☐ date,
- ☐ receiver's full name and complete address,
- ☐ salutation, followed by a colon,
- ☐ complimentary close, followed by a comma,
- ☐ sender's signature and full name,
- ☐ enclosure notation, if necessary;

___ is organized into paragraphs that cover the subject fully in a well-organized way:

- ☐ first paragraph establishes context and states the purpose,
- ☐ middle paragraphs provide all necessary details,
- ☐ last paragraph politely achieves closure;

___ includes no inappropriate content;

___ uses clear, simple language;

___ maintains an appropriate tone, neither too formal nor too conversational;

___ contains no typos or mechanical errors in spelling, capitalization, punctuation, and grammar.

Exercises

■ EXERCISE 2.11

For ten days, save all the business letters you receive. Even though the bulk of them will be junk mail, make a list identifying the *purpose* of each. Prepare a brief oral presentation explaining which letter is the best and which is the worst, and why. (It may be helpful to create overhead transparencies or distribute photocopies to the class, assuming the letters do not contain confidential information.)

■ EXERCISE 2.12

A consumer product that you especially like is suddenly no longer available in retail stores in your area. Write the manufacturer an inquiry letter requesting information about the product and how to place an order.

■ EXERCISE 2.13

Proceeding as if you have received the information requested in Exercise 2.12, write a letter ordering the product.

■ EXERCISE 2.14

Pretend you have received the product ordered in Exercise 2.13, but it is somehow unsatisfactory. Write the manufacturer a claim letter expressing dissatisfaction and requesting an exchange or a refund.

■ EXERCISE 2.15

Team up with a classmate, exchange the claim letters you each wrote in response to Exercise 2.14, and write adjustment letters to each other.

■ EXERCISE 2.16

Write a claim letter expressing dissatisfaction with some product or service that you have actually been disappointed with in the recent past. After the letter has been returned to you with your instructor's corrections, you should then actually mail your claim letter, and see if you receive a reply or perhaps even some form of compensation.

■ EXERCISE 2.17

Write an acknowledgment letter to the editor of either your campus newspaper or a regional daily, expressing your approval of some meaningful contribution made by a local person or organization.

■ EXERCISE 2.18

The writer of this form letter appears to have no knowledge of standard styles of letter layout. Rewrite the letter, adjusting and correcting irregularities.

Centerton High School

100 School Street Centerton, Iowa 50300

January 11, 1999

Dear Classmate,

Remember when the Centerton football team beat City Vocational 7–6 for the County Championship in '88? Or when the Honors Math Club went all the way to the finals in state-wide competition? Or when Mr. Fisk lost his eyeglasses and accidentally went into the women's lavatory at the highway rest area during the class trip? It's hard to believe, but this spring will mark the tenth anniversary of our graduation from good old Centerton High! To celebrate this landmark, a Class of '89 committee—myself included—is working on a special reunion event starting at 6 p.m. on Saturday, May 10, at the Union Hall on Main Street. Husbands, wives, and "dates" are of course welcome in addition to the grads. Cost is $50 per person, which includes buffet dinner, cash bar, reunion T-shirt and a d.j. playing all our favorite songs from the good old days. Mr. Fisk and many of our other teachers—some now retired—are also being invited to attend (free of charge). Please try to make it—the reunion won't be the same without you. You can complete the enclosed pre-registration form indicating your intention to attend, and your T-shirt size. We'd also like payment (or at least a $25/person deposit) at this time. Hope to see you at the reunion!

Yours truly, *Jane Hermanski (Class of '89)*, CHS Guidance Counselor

■ EXERCISE 2.19

The writer of this letter has adopted a highly artificial and pretentious style. Rewrite the letter to convey the message in "plain English."

COUNTY DEPARTMENT OF SOCIAL SERVICES

County Building, Northton, MN 55100

November 10, 1999

Ms. Sally Cramdon
359 Roberts Road
Northton, MN 55100

Dear Ms. Cramdon:

We are in receipt of your pay stubs and your letter of 5 November 1997 and have ascertained a determination re: your application for food stamp eligibility.

Enclosed please find photocopy of food stamp budget sheet prepared by this office on above date, counterindicating eligibility at this point in time. Per County eligibility stipulations, it is our judgement that your level of fiscal solvency exceeds permissible criteria for a household the size of your own (four persons).

In the subsequent event that your remuneration should decrease, and remain at the decreased level for a period of thirty (30) calendar days or more, please do not hesitate to petition this office for a reassessment of your eligibility status at that juncture.

Very truly yours,

Daniel Kuffner

Daniel Kuffner
Casework Aide

■ EXERCISE 2.20

The writer of this letter has committed a great many fundamental blunders, typos, and mechanical errors. Rewrite the letter, fixing all problems.

20/20 Optical Supply, Inc.

North Side Plaza Northweston, WA 98501

August 10, 1899

Service Manger
Northweston Plumbing
23 Reynolds street
Northweston, Wa 98501

Dear Northwesern Plumbing;

Last week your worker's installed a new 50-gallon hot water heater in the basement of are North Side Plazza retail store, now the heater is leaking all over the floor.

Every time I call your phone number I get a recording thet say's you will return my call but you never do. As this has been going on for more than a weak I must ensist that you either call imediatley or send a service person. I'm getting tried of moping up water!!!

Please see to this at your very earlyest convience!

Your's truely

Peter Cooper

Peter Cooper
Store Manger

3

Effective Visuals: Tables, Graphs, Charts, and Illustrations

Learning Objective

Upon completing this chapter, you will be able to enhance your written and oral reports with effective visual elements such as tables, graphs, charts, and pictures.

Much human communication takes place without benefit of actual language—through gestures and facial expressions, for example, and of course by means of diagrams, pictures, and signs. Consider the familiar displays shown here:

Workplace communications make extensive use of visual aids along with text. Proposals, manuals, instructions, and reports of all kinds contain numerous illustrations—often computer-generated—to capture and hold people's attention and help convey information. To function successfully in today's increasingly sophisticated workplace, an employee must be well acquainted with these visual elements. Along with providing a brief overview of basic principles governing their use, this chapter explores the four main categories of visuals—tables, graphs, charts, and illustrations—and explains the principal features and applications of each.

Principles of Effective Visuals

Like good writing, effective visuals are simple, clear, and easy to understand. It is also very important to choose the most appropriate *type* of visual for the task at hand. Today's computer software packages are helpful in this regard, enabling you to assemble data on a spreadsheet, for example, and then to portray it in whatever format is most suitable, or to create eye-catching and informative graphics. Computer technology can produce highly polished results while encouraging a great deal of experimentation with various design features. Like computerized text, graphics stored on disk have the added advantage of easy revision if your data change.

Ironically, the one potential drawback of computer-generated graphics derives from the same versatility that makes these programs so exciting to work with. Inexperienced users can become carried away with the many options at their disposal, creating cluttered, overly elaborate visuals that confuse rather than illustrate. As with writing and page design, simpler is better. Bear always in mind that visuals should never be introduced simply for their own sake, to "decorate" a document. Theoretically, every visual should be able to stand alone, but its true purpose is to clarify the text it accompanies.

When using any kind of visual aid—whether computer-generated or created by less advanced methods—you need to observe the following fundamental rules:

- Number and title every visual in your document sequentially, with outside sources clearly identified. If the document contains only one visual, you can omit the number. Titling a visual is much like writing a subject line for a memo. The title should be brief, accurate, and informative. To write a good title, answer this question: In just a few words, what does this visual depict? The number, title, and source usually appear *beneath* the visual rather than above. (Tables are the exception to this, as they are often numbered and labeled *above*.)
- Any information you provide in a visual you must first discuss in the text, which should then refer the reader to the visual (for example, "See Figure 5"). The visual should be positioned logically, as soon as possible *after* the reference.
- Present all visuals in an appealing manner. They must not be crowded by the text or appear squeezed in, but rather should be surrounded by ample white space.
- Clearly label all elements of the visual and provide a "key" whenever necessary to show scale, direction, and the like. Labels must be easy to read, with their terms matching those used in the text; you cannot call something "x" in the text and label it "y" in the visual if you expect the reader to find it easily.
- When visuals accompany instructions, the point of view in the visuals must be the same as that of the reader performing the illustrated procedure. An overhead view, for example, might confuse if the reader will be approaching the task head-on.
- In no case should a visual omit or manipulate information in order to deceive or mislead. Since the purpose of a visual is to reinforce the meaning of your text, any visual you include is subject to the same strict standards of honesty and accuracy that your text must meet.
- Avoid any spelling mistakes, poor grammar, inconsistent formatting, or other such blunders in the labels, key, title, or other text accompanying a visual. Nothing undermines the credibility of a visual faster than a careless mechanical error.

Tables

The purpose of tables is to portray statistical and other information for easy comparison. Tables consist of horizontal and vertical lines (called "rules") that intersect to create boxes in which to organize and display data. The uppermost horizontal column, which holds the headings of the vertical columns, is called the "boxhead"; the left-hand vertical column, which holds the headings of the horizontal columns, is called the "stub." This arrangement permits ready access to information that would be exceptionally difficult to sort out if it were presented only in text form. A convenient example would be the sports league standings that appear on the sports pages of most newspapers, enabling fans to determine at a glance the won–lost record, percentage, ranking, and other data pertaining to the performance of each team.

Consider as well the following paragraph, which is so full of numerical and other detail as to defy retention:

> According to Hotel Appraisals, Inc., a consulting firm based in Mineola, New York, some of the highest selling prices for major hotels in the United States during the late 1980s were as follows: $251,837,000, in September of 1988, for the 2019-room Hyatt in Chicago; $250,000,000, in July of 1987, for the 482-room La Costa in Carlsbad, California; $129,000,000, in March of 1989, for the 1329-room Marriott in New Orleans; $125,000,000, in December of 1987, for the 470-room Wilshire in Beverly Hills; and $92,000,000, in January of 1989, for the 402-room Stanford in San Francisco.

Certainly this information would be far better presented in table format, as the following example clearly demonstrates.

Highest Selling Prices for Major American Hotels, Late 1980s

Hotel	Location	Rooms	Date	Price
Hyatt	Chicago	2019	9/88	$251,837,000
La Costa	Carlsbad, CA	482	7/87	$250,000,000
Marriott	New Orleans	1329	3/89	$129,000,000
Wilshire	Beverly Hills	470	12/87	$125,000,000
Stanford	San Francisco	402	1/89	$ 92,000,000

Source: Hotel Appraisals, Inc., Mineola, NY.

Sometimes a table will include subdivisions within categories. In such cases, you can employ various design options to avoid confusion. In the following example, the subdivisions are achieved by means of *partial* horizontal rules within the two main "Violation" categories.

Penalties for Selected Driving Offenses in New York

Violation	Mandatory Fine	Maximum Jail Term	Mandatory Action Vs. License
DRIVING WHILE INTOXICATED			
First Violation	$350–$500	1 year	Revoked at least 6 mos.
2 or more violations in 10 years	$500–$5000	4 years	Revoked at least 1 year.
DRIVING WHILE ABILITY IMPAIRED			
First violation	$250	15 days	Suspended 90 days.
2 violations in 5 years	$350–$500	30 days	Revoked at least 6 mos.
3 violations in 10 years	$500–$1500	90 days	Revoked at least 6 mos. if current violation occurred within 5 years of previous violation.

Source: New York State Department of Motor Vehicles.

Graphs

Graphs are used to display statistical trends, changes, and comparisons. There are essentially two kinds: line graphs and bar graphs.

Line Graphs

The primary purpose of line graphs is to portray change over a period of time. A line graph is created by plotting points along horizontal and vertical axes (the x and y axes, respectively), and then joining the points by means of straight lines. The horizontal axis identifies the categories of information that are being compared (the "fixed" or "independent" variables—usually, chronological intervals), while the vertical axis identifies the incremental values that are being compared (the "dependent" variables). Figure 3.1, for example, depicts a company's annual profits during a ten-year period.

Additional lines can be added for purposes of comparison, but each line must appear different to avoid confusion. For example, one line can be solid and another broken, or lines can be drawn in contrasting colors, as in Figure 3.2, which compares the annual profits of two competing

FIGURE 3.1

Line Graph Showing Profits, Company A, 1990 to 1999

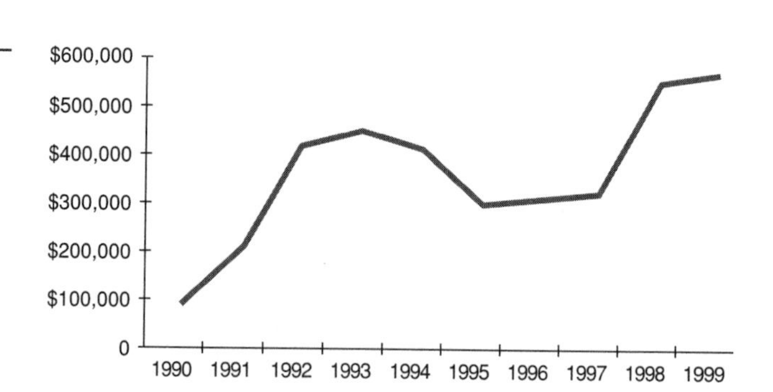

FIGURE 3.2

Line Graph Showing Profits, Company A and Company B, 1990 to 1999

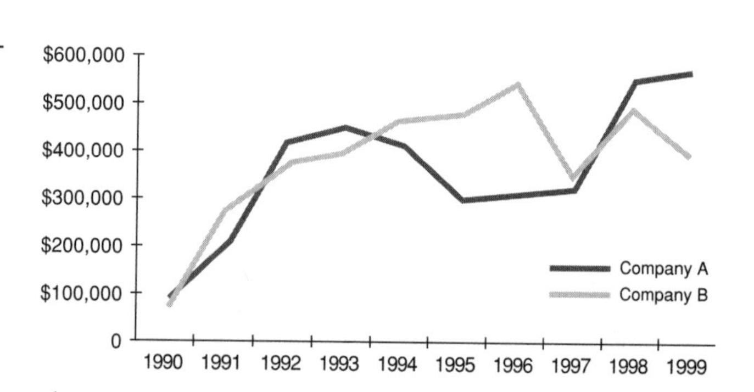

companies during a ten-year period. Note the key, which indicates that the darker line represents Company A, while the lighter line represents Company B.

Bar Graphs

Like tables and line graphs, bar graphs are very useful for comparing data. As do line graphs, bar graphs consist of horizontal and vertical axes that list the dependent and independent variables, but this is determined by whether the bars themselves are horizontal or vertical. If the bars are vertical, the vertical axis will list the dependent variables, while the horizontal axis will list the independent variables. Figure 3.3, for example, is a vertical bar graph that portrays the gradual decline in average per capita beef consumption from 1970 to 1994.

If the bars are horizontal, the arrangement is reversed, with the horizontal axis listing the dependent variables and the vertical axis listing the independent variables. Horizontal bar graphs are useful for accommodating large numbers of bars and for portraying consistently increasing or decreasing values. For example, you might be impressed to read that "state lottery proceeds increased more than tenfold during the period from 1980 to 1995," but such information is even more striking when graphically displayed, as in Figure 3.4.

Horizontal bars offer the added advantage of permitting the independent variables to be labeled horizontally if those labels are comparatively lengthy. See, for example, Figure 3.5, in which this feature is especially helpful. To create comparisons within categories of information, each bar can be presented alongside an accompanying bar or two, but the additional bar(s) must be drawn differently to avoid confusion. Figure 3.5 illustrates participation by men and women in the ten most popular sports activities in 1994. Note the key, which indicates that the darker bars represent female participation, while the lighter bars represent male participation.

FIGURE 3.3

Bar Graph Showing Average Per Capita Beef Consumption, 1970 to 1994

Source: Statistical Abstract of the United States, 1996, p. 108.

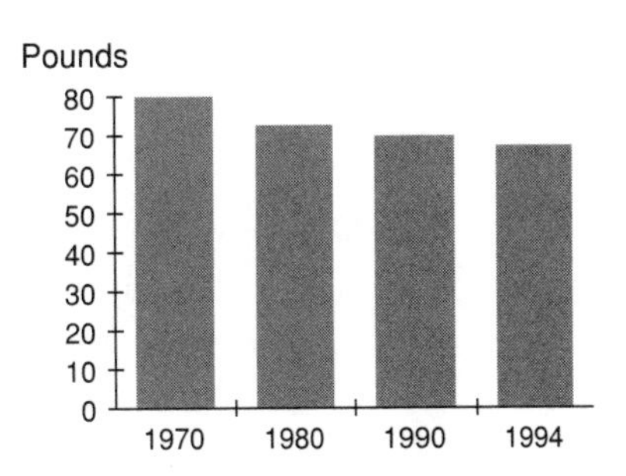

FIGURE 3.4

Bar Graph Showing State Lottery Proceeds, 1980 to 1995

Source: Statistical Abstract of the United States, 1996, p. 296.

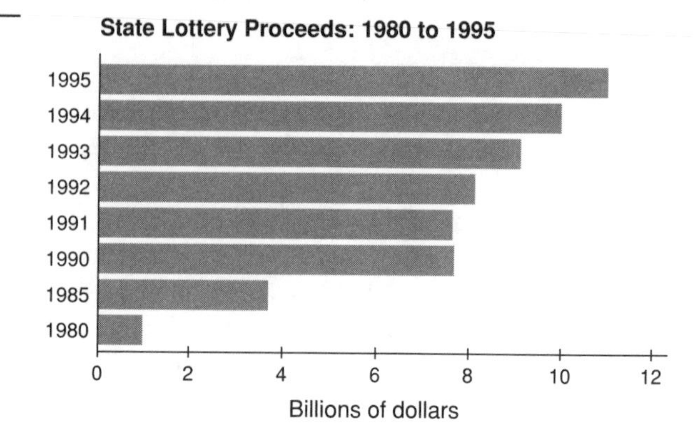

FIGURE 3.5

Bar Graph Showing Most Popular Sports Activities, by Sex, 1994

Source: Statistical Abstract of the United States, 1996, p. 248.

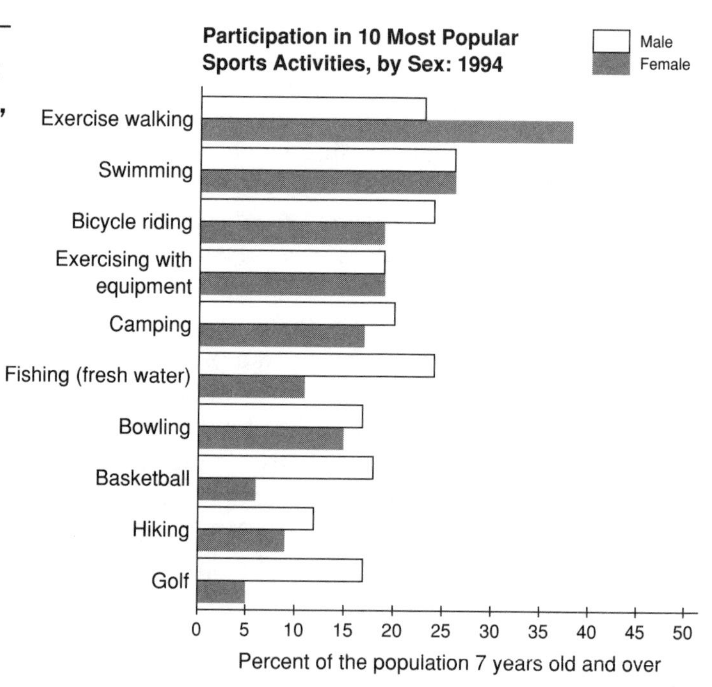

Charts

The purpose of a chart is to portray quantitative, cause-and-effect, and other relationships among the component parts of a unified whole. Comprising squares, rectangles, triangles, circles, and other geometric shapes linked by plain or arrowhead lines, charts can depict the steps in a production process, the chain of command in an organization, and other sequential or hierarchical interactions. Among the principal kinds of charts are flow charts, organizational charts, and circle charts.

Flow Charts

A flow chart is used typically to portray the steps through which work (or a process) must "flow" in order to reach completion. The chart clearly labels each step, and arrows indicate the sequence of the steps so that someone unfamiliar with the process can easily follow it. Flow charts are usually read from top to bottom or from left to right, although some depict a circular flow. The chart in Figure 3.6, for example, shows how a successful bill is signed into law.

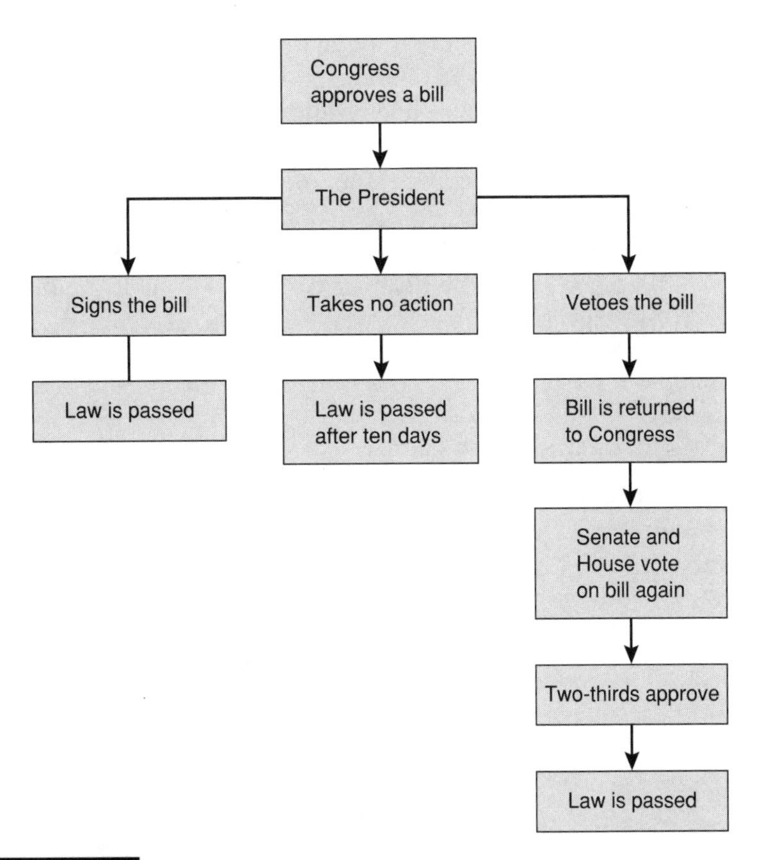

FIGURE 3.6 Flow Chart Showing How a Bill Is Signed Into Law

Organizational Charts

Like flow charts, organizational charts consist of labeled boxes linked by lines or arrows. Organizational charts portray chains of command within businesses, agencies, and other collective bodies, indicating who has authority over whom, and also suggesting the relationships among various functional areas or components within the organization. Not surprisingly, the most powerful positions are placed at the top, the least powerful at the bottom. Those on the same horizontal level are at approximately equal levels of responsibility. Figure 3.7, for example, shows the newsroom organization of a small daily newspaper.

Circle Charts

Among the most familiar of all visual devices, circle—or "pie"—charts are often used to show the percentage distribution of money. As such, they are useful in analyzing relative costs and profits. Their more general application is simply to depict relationships among parts within statistical wholes. In that broader context, they facilitate such tasks as risk analysis, needs assessment, and resource allocation.

Each segment of a circle chart resembles a slice of pie and constitutes a percentage. For maximum effectiveness, the pie should include at least three but no more than seven slices. (To limit the number of slices without omitting data, several small percentages can be lumped together under the heading of "Other.") As if the pie were a clock face, the biggest slice begins at 12 o'clock, and the slices get progressively

FIGURE 3.7

Organizational Chart Showing Newsroom Organization of Small Daily Newspaper

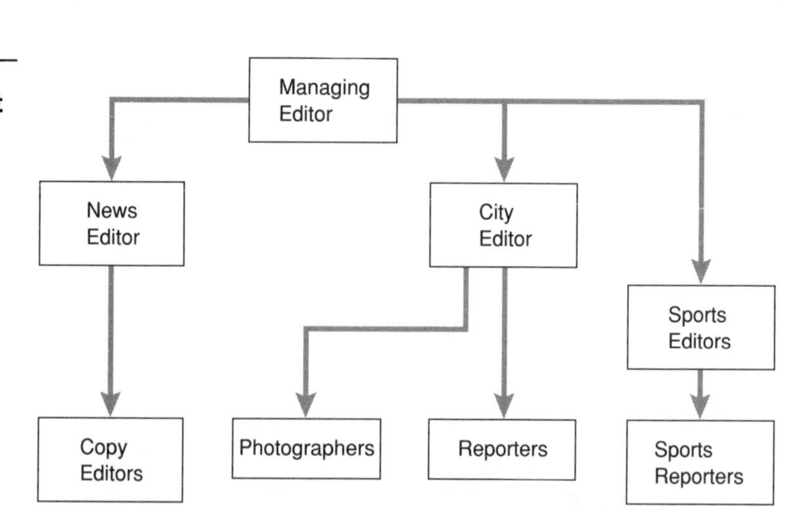

FIGURE 3.8

Circle Chart Showing Cost of Attending Residential Community College, One Semester

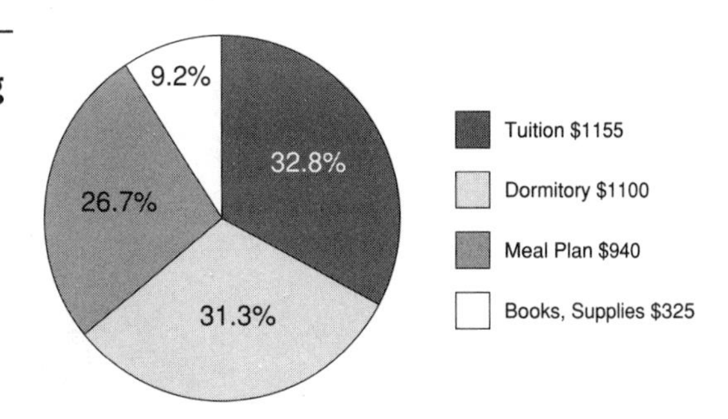

smaller as they continue, clockwise, around the circle. If colored or shaded, the slices tend to get lighter as they continue around. Each slice is labeled, showing its percentage of the total. (Obviously, the slices must add up to 100 percent.) A key must be provided, to identify what each slice represents. Figure 3.8 shows the costs of attending a residential community college for one semester.

Illustrations

Illustrations—be they photographs, drawings, or diagrams—are another highly effective form of visual aid. Each type has certain advantages. As with so many aspects of workplace communications, which you choose will depend upon the purpose and audience of your communication.

Photographs

A photograph is an exact representation; its main virtue, therefore, is strict accuracy. Indeed, photos are often required in certain kinds of documents, such as licenses, passports, accident reports (especially for insurance purposes), and patent applications. Photos are often used in law enforcement, whether to warn the public of criminals in "Most Wanted" posters depicting the faces of fugitives, or to document the scene of a crime or accident. Figure 3.9, for example, is a photograph documenting vehicle damage following a collision.

Ideally, photos should be taken by trained professionals. Even an amateur, however, can create reasonably useful photos by observing the following fundamental guidelines:

- Use a good 35 mm camera with an adjustable lens.
- Use black-and-white film, which produces sharper contrast.

FIGURE 3.9

Photograph Documenting Damaged Vehicle

Charles Kershner

- Ensure that the light source, whether natural (the sun) or artificial (floodlamp or other electrically generated light) is behind you; avoid shooting *into* the light.
- Stand close enough to your subject to eliminate surroundings unless relevant.
- Try to focus on the most significant part of your subject to minimize unwanted detail. (If you have access to darkroom facilities, unwanted detail can be cropped out and the remaining image enlarged.)
- To provide a sense of scale in photographs of unfamiliar objects, include a familiar object within the picture. In photos of small objects, for example, a coin or paper clip works nicely. In photos of very large objects, a human figure is helpful.
- Hold the camera absolutely still while taking the picture. If possible, mount the camera on a tripod and use an automatic shutter release.

Drawings

The purpose of most drawings—whether freehand or computer-assisted—is to create clear, realistic depictions of objects under discussion. The main advantage is that in a drawing you can easily omit unwanted detail and portray only what is most relevant. In addition, a drawing can distort information it depicts in the interest of clarity, by simplifying, enlarging, or otherwise emphasizing key features. One obvious example of this is a type of drawing called a floorplan, which—like a map—is certainly much clearer and more informative than an overhead photo would be (see Figure 3.10).

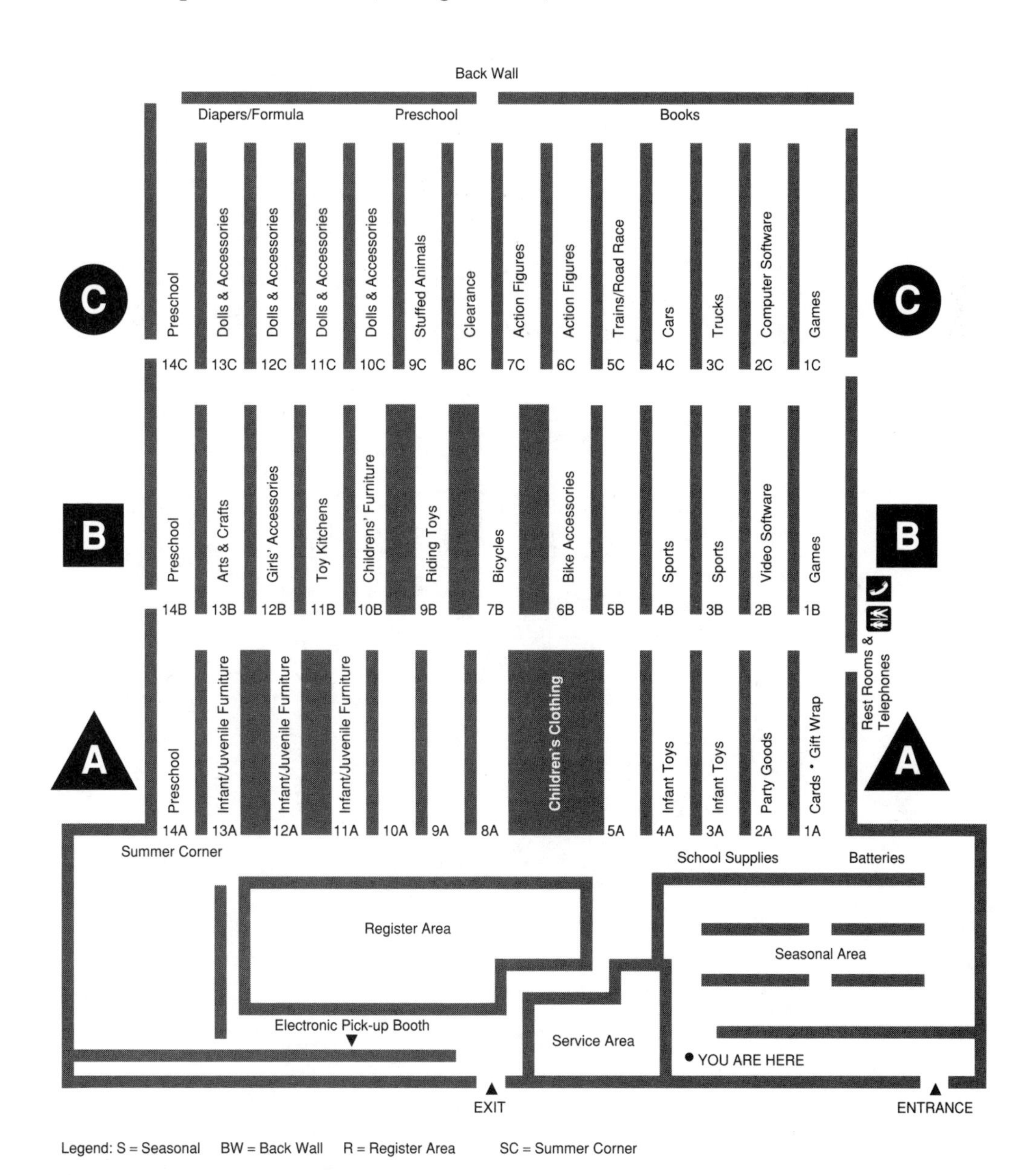

FIGURE 3.10 Floorplan of Toys Я Us

Source: Toys Я Us Store Directory, 1995.

Other useful applications are the "exploded" drawing so often used in product assembly instructions (see Figure 3.11) and the cross-section, or "cutaway view," which provides visual access to the interior workings of mechanisms and other objects (see Figure 3.12). Moreover, drawings can be combined with tables, graphs, and charts to create pictographs that enliven otherwise routine documents. See, for example, Figure 3.13, an organizational chart in which desks are a creative pictorial substitute for the conventional boxes.

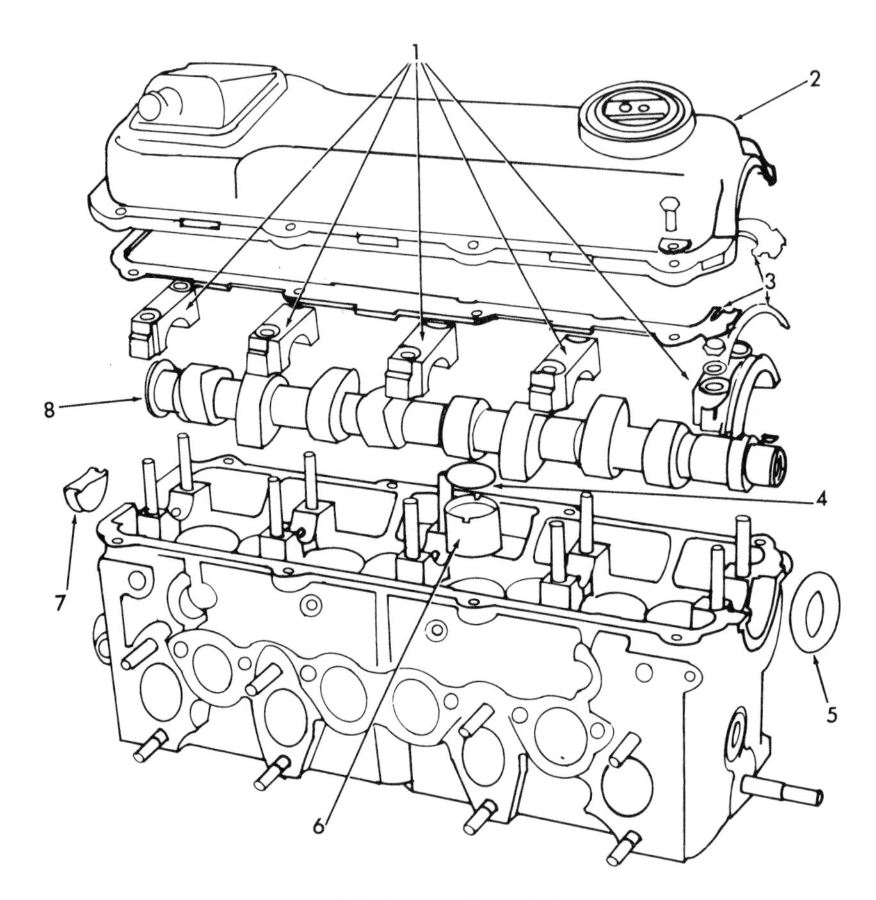

Exploded view of camshaft assembly

1. Camshaft bearing caps
2. Camshaft cover
3. Gasket
4. Valve adjusting disc
5. Oil seal
6. Cam follower
7. End plug
8. Camshaft

FIGURE 3.11 Exploded View of Volkswagen Rabbit Camshaft Assembly

Source: Chilton's Rabbit/Scirocco Repair & Tune-Up Guide. Radnor, PA: Chilton, 1975, p. 51.

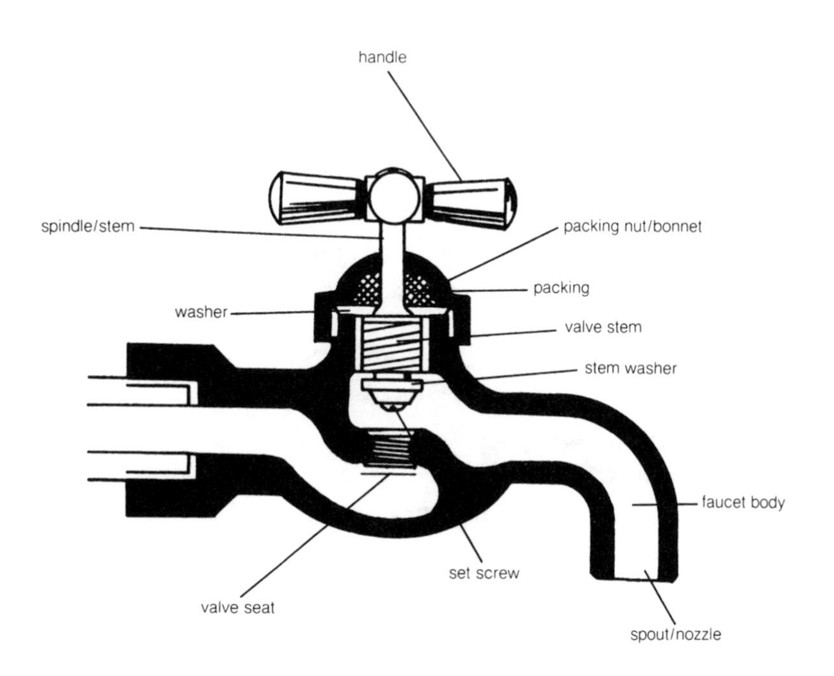

FIGURE 3.12 **Cutaway View of Faucet**

Source: Bragonier, R., and David Fisher. *What's What: A Visual Glossary of the Physical World.* Maplewood, NJ: Hammond, 1981, p. 269.

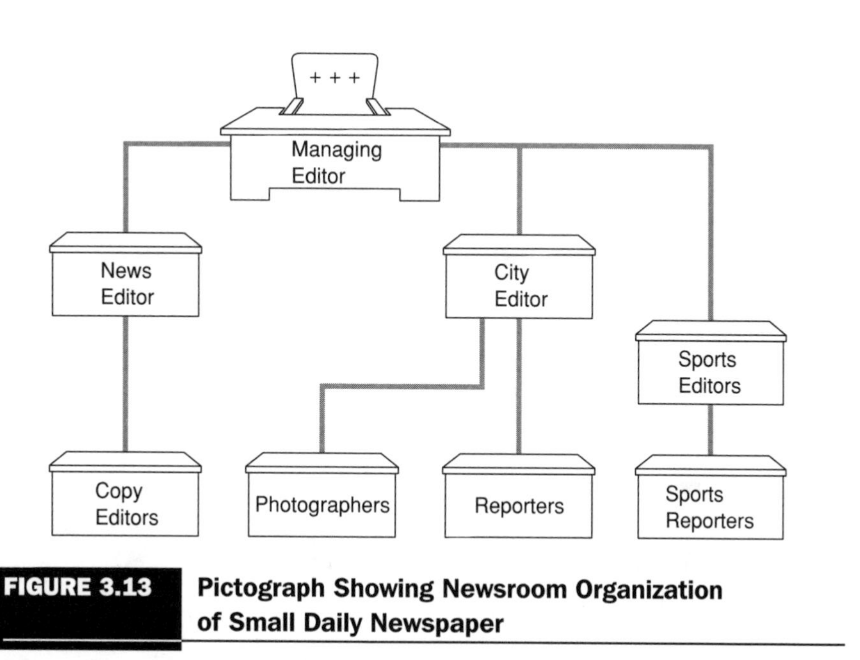

FIGURE 3.13 **Pictograph Showing Newsroom Organization of Small Daily Newspaper**

Source: Brooks, Brian, et al. *News Reporting and Writing.* New York: St. Martin's Press, 1992, p. 28.

FIGURE 3.14

Clip Art of Computer Cartoon and Letter Cartoon

Source: T. Maker Co., 1988.

Another popular source for images in communications is "clip art," which are collections of prepackaged illustrations that span a vast range of topics. Clip art can be purchased independently or as integral parts of computer programs for word processing. You risk trivializing your work, however, unless you exercise selectivity when using clip art; much of it is excessively whimsical, as Figure 3.14 shows.

Diagrams

Just as a drawing can be considered a simplified photograph, a diagram can be considered a simplified drawing. In Figure 3.15, for example, we see how a three-way switch is wired, even though this is not in fact what the wiring looks like. Yet anyone conversant with electrical symbols can actually read the drawing more easily than the realistic—and far more complex—rendering in Figure 3.16. Most diagrams—blueprints, for example, or engineering graphics—require advanced familiarity on the reader's part, and are therefore useful only in documents intended for technicians and other specialists.

FIGURE 3.15

Schematic Drawing of Three-Way Switches

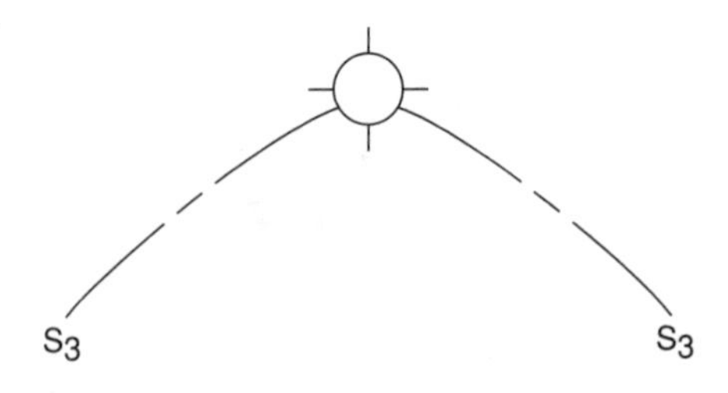

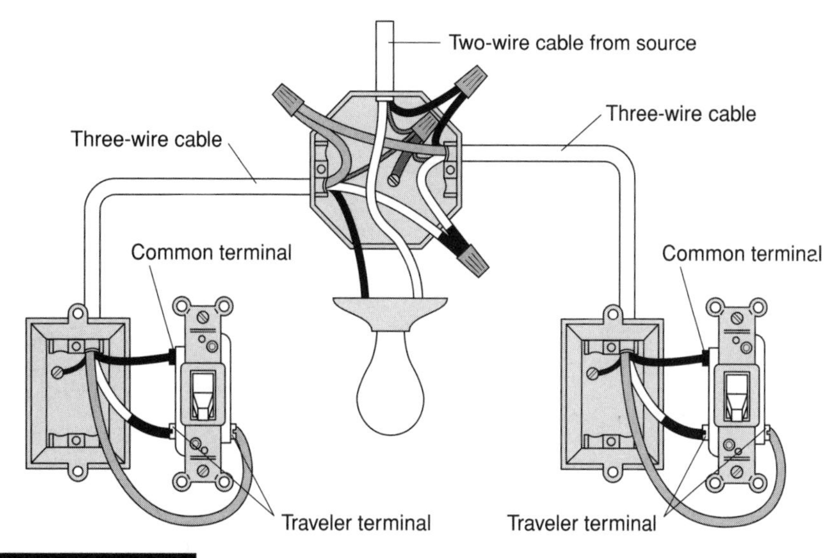

FIGURE 3.16 **Realistic Rendering of Three-Way Switches**

Source: Reader's Digest New Complete Do-It-Yourself Manual. Pleasantville, NY: Reader's Digest Association, 1991, p. 253.

☑ Checklist Evaluating a Visual

A good visual

___ is the most appropriate choice—table, graph, chart, or picture—for a particular communication;

___ is numbered and titled, with the source (if any) identified;

___ occupies the best possible position within the document, immediately after the text it clarifies;

___ does not appear crowded, with enough white space surrounding to ensure effective page design;

___ includes clear, accurate labels that plainly identify all elements;

___ includes a key if necessary for further clarification;

___ maintains consistency with all relevant text in terms of wording, point of view, and so on;

___ upholds strict standards of accuracy;

___ contains no typos or mechanical errors in spelling, capitalization, punctuation, and grammar.

Exercises

EXERCISE 3.1

Choosing from Column B, identify the most appropriate kind of visual for depicting each item in Column A.

Column A	Column B
Registration procedure at your college	table
Interest-rate fluctuations during the past ten years	photograph
Inner workings of a steam boiler	line graph
Structure of the U.S. Executive Branch	bar graph
Average salaries of six selected occupations	cutaway view
Uniform numbers, names, ages, hometowns, heights/weights, and playing positions of the members of a college football team	diagram
House for sale	flow chart
Percentage distribution of your college's student body by major	exploded view
Automobile steering mechanism	organizational chart
Circuitry of an electronic calculator	pie chart

EXERCISE 3.2

Create a table showing the current cost per gallon for regular and high-octane of three major brands of gasoline. Include self-serve and full-service variables, if applicable.

EXERCISE 3.3

Create a line graph showing your favorite professional sports team's place in the league standings for the past ten seasons.

EXERCISE 3.4

Building upon the information in Exercise 3.3, create a double-line graph comparing your favorite professional sports team's place in the league standings for the past ten seasons to that of one other team.

■ EXERCISE 3.5

Create a bar graph showing the total population of the New England states.

■ EXERCISE 3.6

Write a two- or three-page report explaining a process related to your field of study or employment. Include a flow chart depicting that process.

■ EXERCISE 3.7

Write a two- or three-page report about a club or other group to which you belong. Include a chart showing its organizational structure.

■ EXERCISE 3.8

Create a circle graph showing how you spend your money in a typical month.

■ EXERCISE 3.9

Create drawings of any three of the following: a flashlight (cutaway), an electric plug (exploded), the route from your home to your college (map), three different kinds of hammer, and the floorplan of a place where you have worked.

■ EXERCISE 3.10

Find and photocopy or otherwise reproduce an example of each kind of visual discussed in this chapter. Write a booklet report that incorporates the visuals and evaluates them for clarity and effectiveness.

4

Short Reports: Page Design, Formats, and Types

Learning Objective Upon completing this chapter, you will be able to apply the basic principles of page design and format to write effective short reports of various kinds.

Like memos and letters, reports are a major form of on-the-job communication, can be internal or external documents, and follow certain standard conventions. In several major respects, however, reports are quite different from memos and letters.

For example, a report is rarely only a written account of information the reader already has. Nearly always, the report's subject matter is new information. The reader may be acquainted with the general outline of the situation the report explores, but not with the details. Very often, in fact, the reader will have specifically requested the report in order to get those particulars. Indeed, reports exist for that very purpose, to communicate needed information that is too complicated for a memo or letter. Stated in the simplest terms, there are essentially two kinds of reports: short and long. This chapter focuses on the former, discussing basic principles of page design, short report formats, and several common types of short reports.

Page Design

As we have seen, the physical characteristics of memos and letters are largely determined by established guidelines that vary only slightly. But reports, while also subject to certain conventions, are to a much greater degree the creation of individual writers who determine not only their content but also their physical appearance. This is significant because our ability to comprehend what we read is greatly influenced by its arrangement on the page. A report, therefore, should never *look* difficult or intimidating. Consider, for example, Figure 4.1, which has been adapted from a safety manual for railroad employees.

Obviously, the passage is nearly unreadable in its present state. To make it more visually appealing, begin by inserting more space between the lines, and by using a slightly larger type size and both uppercase and lowercase letters, as shown in Figure 4.2.

Certainly the revised page is far more legible. It can be improved still further, however, by organizing the content into paragraphs, and adopting a ragged right-hand margin (see Figure 4.3).

The use of varied spacing, lists, and boldface headings, as well as some minor editing, will make the content emerge even more clearly.

Obviously, Figure 4.4 is easier to read than the earlier versions. Certainly such revision is worthwhile, and not particularly difficult if the following fundamental principles of effective page design are observed.

ELECTRIC SHOCK

ELECTRIC SHOCK IS NOT ALWAYS FATAL, AND RARELY IS IT IMMEDIATELY FATAL. IT MAY ONLY STUN THE VICTIM AND MOMENTARILY ARREST BREATHING. IN CASES OF ELECTRIC SHOCK, BREAK CONTACT, RESTORE THE VICTIM'S BREATHING BY MEANS OF ARTIFICIAL RESPIRATION, AND MAINTAIN WARMTH. TO AVOID RECEIVING A SHOCK YOURSELF, EXERCISE EXTREME CAUTION WHEN ATTEMPTING TO RELEASE THE VICTIM FROM CONTACT WITH A LIVE CONDUCTOR. MANY PERSONS, BY THEIR LACK OF KNOWLEDGE OF SUCH MATTERS, HAVE BEEN SEVERELY SHOCKED OR BURNED WHEN ATTEMPTING TO RESCUE A CO-WORKER. TO RELEASE A VICTIM FROM CONTACT WITH LIVE CONDUCTORS KNOWN TO BE 750 VOLTS OR LESS, DO NOT TOUCH THE CONDUCTOR, AND DO NOT TOUCH THE VICTIM OR THE VICTIM'S BARE SKIN IF THE VICTIM IS IN CONTACT WITH THE LIVE CONDUCTOR. INSTEAD, USE A PIECE OF DRY, NON-CONDUCTING MATERIAL SUCH AS A PIECE OF WOOD, ROPE, OR RUBBER HOSE TO PUSH OR PULL THE LIVE CONDUCTOR AWAY FROM THE VICTIM. THE LIVE CONDUCTOR CAN ALSO BE HANDLED SAFELY WITH RUBBER GLOVES. IF THE VICTIM'S CLOTHES ARE DRY, THE VICTIM CAN BE DRAGGED AWAY FROM THE LIVE CONDUCTOR BY GRASPING THE CLOTHES—NOT THE BARE SKIN. IN SO DOING, THE RESCUER SHOULD STAND ON A DRY BOARD AND USE ONLY ONE HAND. DO NOT STAND IN A PUDDLE OR ON DAMP OR WET GROUND. TO RELEASE A VICTIM FROM CONTACT WITH LIVE CONDUCTORS OF UNKNOWN VOLTAGE OR MORE THAN 750 VOLTS . . . [text continues]

FIGURE 4.1 Poor Page Design

- ***Legible Type.*** Although many different typefaces and type sizes exist, most readers respond best to twelve-point type using both uppercase and lowercase letters, like this text. Anything smaller or larger is difficult to read, as is the all-capitals approach; such options are useful only in major headings or to emphasize a particular word or phrase.

ELECTRIC SHOCK

Electric shock is not always fatal, and rarely is it immediately fatal. It may only stun the victim and momentarily arrest breathing. In cases of electric shock, break contact, restore the victim's breathing by means of artificial respiration, and maintain warmth. To avoid receiving a shock yourself, exercise extreme caution when attempting to release the victim from contact with a live conductor. Many persons, by their lack of knowledge of such matters, have been severely shocked or burned when attempting to rescue a co-worker. To release a victim from contact with live conductors known to be 750 volts or less, do not touch the conductor, and do not touch the victim or the victim's bare skin if the victim is in contact with the live conductor. Instead, use a piece of dry, non-conducting material such as a piece of wood, rope, or rubber hose to push or pull the live conductor away from the victim. The live conductor can also be handled safely with rubber gloves. If the victim's clothes are dry, the victim can be dragged away from the live conductor by grasping the clothes—not the bare skin. In so doing, the rescuer should stand on a dry board and use only one hand. Do not stand in a puddle or on damp or wet ground. To release a victim from contact with live conductors of unknown voltage or more than 750 volts . . . .

FIGURE 4.2 Revised Page

- ***Generous Margins.*** Text should be centered on the page and framed by white space. The top and bottom margins should both be at least 1 inch and the side margins 1.25 inches. If the report is to be bound, the left margin should be 2 inches. (If the report is to be duplicated back-to-back before binding, then the 2-inch margin should be on the *right*-hand side of the even-numbered pages.) The right-hand margin should not be justified, but should run ragged; this improves legibility by creating greater contrast from line to line.

ELECTRIC SHOCK

Electric shock is not always fatal, and rarely is it immediately fatal. It may only stun the victim and momentarily arrest breathing. In cases of electric shock, break contact, restore the victim's breathing by means of artificial respiration, and maintain warmth. To avoid receiving a shock yourself, exercise extreme caution when attempting to release the victim from contact with a live conductor. Many persons, by their lack of knowledge of such matters, have been severely shocked or burned when attempting to rescue a co-worker.

To release a victim from contact with live conductors known to be 750 volts or less, do not touch the conductor, and do not touch the victim or the victim's bare skin if the victim is in contact with the live conductor. Instead, use a piece of dry, non-conducting material such as a piece of wood, rope, or rubber hose to push or pull the live conductor away from the victim. The live conductor can also be handled safely with rubber gloves. If the victim's clothes are dry, the victim can be dragged away from the live conductor by grasping the clothes—not the bare skin. In so doing, the rescuer should stand on a dry board and use only one hand. Do not stand in a puddle or on damp or wet ground.

To release a victim from contact with live conductors of unknown voltage or more than 750 volts . . . .

FIGURE 4.3 Second Revision

- ***Textual Divisions.*** Long, unbroken passages of text are very difficult to follow with attention, which is why the practice of dividing text into paragraphs was adopted centuries ago. In most workplace writing, paragraphs should not exceed five or six sentences, and should be plainly separated by ample white space. If the paragraphs are single-spaced, insert double-spacing between them; if the paragraphs are double-spaced, use triple-spacing between them. To further organize content, group related paragraphs within a

ELECTRIC SHOCK

Electric shock is not always fatal, and is rarely immediately fatal. It may only stun the victim and momentarily arrest breathing. In cases of electric shock, do three things:

1. Break contact;
2. Restore breathing by artificial respiration;
3. Maintain warmth.

To avoid receiving a shock yourself, exercise extreme caution when attempting to release the victim from contact with a live conductor. Many persons, lacking knowledge of such matters, have been severely shocked or burned attempting to rescue a co-worker.

Release of victim from contact with live conductors known to be 750 volts or less:

- Do not touch the live conductor.
- Do not touch the victim or the victim's bare skin while the victim is in contact with the live conductor.
- Instead, use a piece of DRY non-conducting material such as a piece of wood, rope, or rubber hose to push or pull the live conductor away from the victim. The live conductor may be handled safely with rubber gloves.
- If the victim's clothes are dry, the victim can be dragged away from the live conductor by grasping the clothing—not the bare skin. In so doing, the rescuer should stand on a dry board and . . . .

FIGURE 4.4 Third Revision

report into separate sections that logically reflect the internal organization of the report's information. Like the individual paragraphs, these sections should be plainly separated by proportionately greater spacing.

- ***Headings.*** Separate sections of text should be labeled with meaningful headings that further clarify content and allow the reader to skim the report for specific aspects of its subject matter. Ordinarily, a heading consists of a word or phrase, *not* a complete sentence. The position of a heading is determined by its relative importance. A major heading is set in boldface caps and centered,

LIKE THIS

A secondary heading is set either in uppercase letters or in both uppercase and lowercase, is flush with the margin, and can be underlined or set in boldface print (although not both),

Like This

or

Like This

A subtopic heading is run in to the text, separated by a period or a colon, and is sometimes indented. Set in both uppercase and lowercase letters, it can be underlined or set in bold print (though not both),

Like This. These recommendations are based on those in *The Gregg Reference Manual,* considered to be the most widely recognized authority on such matters.

Obviously, these principles are flexible, and various approaches to heading design and placement are used, some of them quite elaborate. Among the most helpful recommendations in *Gregg* is to limit a report to no more than three levels of heads.

- ***Lists.*** Sometimes a list is more effective than a conventional paragraph. If the purpose of the list is to indicate a definite order of importance, the items in the list should be *numbered* in descending order, with the most important item first, least important last. Similarly, if the list's purpose is to indicate a chronological sequence of events or actions (as in a procedures manual), the items should be numbered in sequential order. Numbers are not necessary, however, in a list of approximately equal items. In those cases, "bullets" (solid black dots, like those used in this section), asterisks, or dashes will suffice.

Thanks to computerized word processing, nearly every workplace writer now has access to many text design features that in the past were available only through commercial printshops. As we have already seen (in the "Electric Shock" example and in the discussion of headings), options such as varied spacing and type size, boldface print, capitalization, and underlining can make your writing appear much more professional.

Used selectively, these features enhance the design of a page not only by signaling major divisions and subdivisions within the content, but also by creating emphasis through the highlighting of key words, concepts, and other text elements. In addition, many software packages are equipped with ready-made report templates that offer additional features such as headers and automatic page numbering for multi-page documents, and these can be adapted to the individual writer's needs.

Remember to exercise restraint and maintain consistency in using any of these tools, however, and to keep your page design relatively simple. It is very easy to get carried away, thereby creating a messy and confusing effect, especially if you are still relatively inexperienced with this technology. Like visual elements, page design options should never be used simply as decoration, but as aids to your reader's understanding. The key is to experiment with your software and thoroughly familiarize yourself with its capabilities. Soon you will develop a more accurate sense of which page design features might be genuinely helpful to your reader.

Report Formats: Memo, Letter, and Booklet

Many companies and organizations still prepare short reports using the old-fashioned, fill-in-the-blanks approach typified by the forms reproduced in Figures 4.5 and 4.6. As we have seen, however, computer technology now enables individual writers to personally design the pages of their reports. Most such customized reports can be categorized according to one of three report formats: memo, letter, or booklet.

Typically used for in-house purposes, the **memo report** is similar to the conventional memo but is longer (two pages or more), and is therefore divided into separate, labeled sections. The **letter report**—typically sent to an outside reader—is formatted like a conventional business letter, except that the letter is divided into separate, labeled sections, much like a memo report. The **booklet report** resembles a short term paper and includes a title page. It too is divided into separate, labeled sections. It is also accompanied by a cover memo (for in-house reports) or cover letter (for reports sent to outside readers). Much like the opening paragraph of a memo report or letter report, this cover document serves to orient the reader by establishing context and explaining the purpose and scope of the booklet report.

EMPLOYEE ACCIDENT/FORM A

TO BE COMPLETED BY EMPLOYEE

Name ________________ Home address ________________

Social security no. ________________ Date of birth ________________

Sex __M__F__ Department in which you work ________________

Accident date ________ Day of week ________ Time _____a.m. _____p.m.

Date accident was reported ________ To whom ________________

Location of accident ________________ Witnesses ________________

Description of accident (what was employee doing, what equipment was being used, etc.)

____

____

____

Description of injury (include nature of injury and body part)

____

____

____

Did you receive medical care on premises? ________ Describe ________________

____

____

If employee is being treated:
Name and address of physician: ________________

Name and address of hospital: ________________

Do you have a second job? ________________

EMPLOYEE SIGNATURE ________________ Date: ________

TO BE COMPLETED BY COMPANY NURSE

Above employee came to me on ________ regarding the above injury.

Comments:

____

____

____

NURSE'S SIGNATURE ________________ Date: ________

FIGURE 4.5 Employee Accident Form

NAME LAST FIRST INITIAL

SOC. SEC. #

OCCUPATIONAL INJURY

MEDICAL SERVICE REPORT

DATE OF REPORT ______

ADDRESS NO. STREET

CITY STATE ZIP

PERM. NO.	Age	SMDSW 1 2 3 4 5	SHIFT 1 2 3	M F 1 2	SERVICE DATE		DISP. 1 2 3	DATE FOR FOLLOW-UP IF CHECKED BELOW	
DATE	DAY	TIME	AM / PM		A	MINOR INJURY		A	DIATHERMY
					B	COMP. INJURY		B	INFRA-RED
					C	LOST TIME INJ.		C	MEDCO-SONOLATOR
								D	ULTRA-VIOLET
								E	WHIRLPOOL
					1	LOCATION		F	X-RAY
					2	AGENCY		G	1st DRESSING
					3	TYPE		H	RE-DRESSING
					4	UNSAFE CONDITIONS		II	CONSULTATION
					5	UNSAFE ACT		J	TETANUS SHOT
					6	PERSONAL FACTOR		K	
					7			1	SENT TO HOSPITAL
					8	DERMATITIS		2	SENT HOME
					9	SYSTEMIC		3	SENT FOR TREATMENT
								4	SENT FOR CONSULTATION
								5	SENT TO OWN M.D.

DEPT. NO.

JOB Pc. WORK Y N — Present

Injured

FOREMAN Notified Y N — Left Work

Probable Return

PLACE OF ACCIDENT: On Job ( ) Other (Describe) — Returned

NATURE OF INJURY — Days Lost

DESCRIPTION OF ACCIDENT

MEDICAL OFFICE - RC SIGNATURE

MACHINE OR TOOL INVOLVED?

WHO SAW ACCIDENT?

WAS MACH. OR TOOL DEFECTIVE/REPORTED?

Y N Y N

WHAT PERSONAL PROTECTIVE EQUIP. USED AT TIME OF ACCIDENT?

Safety Glasses Y N Hearing Protection Y N Goggles Y N OTHER (Describe)

White - Medical Office Canary - Comp. File Pink - Return Follow-up Goldenrod - Safety Dept. NEG. 277

FIGURE 4.6 Occupational Injury Form

Both memo reports and booklet reports often contain visuals; letter reports sometimes do. Figures 4.7, 4.8, and 4.9 illustrate the three formats. Written by a fictitious health inspector and his supervisor, these examples use easily understood subject matter. These three formats can be adapted to any workplace situation simply by changing the headings to suit the context at hand.

Monroe County Health Department

MEMORANDUM

DATE: February 2, 1999

TO: Marjorie Witkowski, Supervisor

FROM: Richard Vaughan, Senior Inspector

SUBJECT: Restaurant Inspections

As you requested, here are the results of last week's inspections of food service establishments in the county, along with a week-by-week statistical summary of inspections during January.

UNSATISFACTORY

The following establishments were found to be in substantial violation of the sanitary code.

Big Daddy's Steak House
431 Grand Avenue, Conover Falls
Inspected January 25, 1999

Toxic chemicals (antifreeze, can of ant/roach killer) found on premises. Potentially hazardous foods not kept at or above 140 degrees F during hot holding. Food not protected—buckets of food stored on floor in cooler, food not

FIGURE 4.7 **Memo Report, page 1**

2

covered in coolers. Raw meat stored over prepared foods in cooler. Food build-up in storage room refrigerator. Canned goods in poor condition (dented, rusted). Bowl used as flour scoop. Box of paper towels improperly stored on floor. Non-food contact surfaces not easily cleanable. Cardboard used as liner on food storage shelves. Restroom missing hand wash sign. Light fixture missing shield and end caps. Kitchen ceiling tiles missing. No 1999 permit on display.

Employee Cafeteria, Paragon Insurance Co.
Airport Road, Cedarville
Inspected January 26, 1999

Potentially hazardous foods not kept at or below 45 degrees F during cold holding. Potentially hazardous foods not kept at or above 140 degrees F during hot holding. Single-service napkins stored on kitchen floor.

Roma Pizzeria
38 Crowley Street, Dunkirk
Inspected January 27, 1999

Worker serving pizza slices with bare hands. Potentially hazardous foods not kept at or above 140 degrees F during hot holding. Food not protected—uncovered food in freezer, salt bucket not labeled. Hair improperly restrained—hats, nets/visors required. In-use utensils stored on paper plate. Employee (delivery driver) smoking in kitchen.

SATISFACTORY

The following establishments were found to be in essential compliance with the sanitary code, although some violations were noted.

Imperial Wok
618 Rogers Street, Cooperton
Inspected January 28, 1999

Potentially hazardous foods not kept at or above 140 degrees F during hot holding. Food not protected—jars of juice stored on kitchen floor. Improper use of utensils—scoop stored handle down in flour.

FIGURE 4.7 Memo Report, page 2

3

Cuzzie's Pub
39 Railroad Street, Monroe
Inspected January 29, 1999

Unshielded light fixture in walk-in cooler. No hand soap in restroom.

NO VIOLATIONS

The following establishments were found to be in full compliance with the sanitary code.

Conover Falls Coffee House
17 Village Green East, Conover Falls
Inspected January 25, 1999

Mister Eight Ball
49 Clinton Street, Dunkirk
Inspected January 27, 1999

SUMMARY OF JANUARY 1999 INSPECTIONS

	Unsatisfactory	Satisfactory	No Violations
Jan. 4–8	3	2	2
Jan. 11–15	5	0	1
Jan. 18–22	2	3	2
Jan. 25–29	3	2	2
Totals	13	7	7

FIGURE 4.7 Memo Report, page 3

Monroe County Health Department

County Office Building ✦ Court House Square
Monroe, Wyoming 82001 ✦ (307) 555-1200

February 5, 1999

Mr. Daniel Runninghorse, Editor
The Monroe Daily Observer
687 Harpur Street
Monroe, WY 82001

Dear Mr. Runninghorse:

As you may know, the County Health Department conducts ongoing, unannounced inspections of food service establishments to ensure their compliance with state codes, rules, and regulations. Since the findings of these inspections are a matter of public record, The Monroe Daily Observer has in the past printed that information in its entirety. Now that you have become the new Editor of the Observer, we would like you to continue this practice, which we regard as a valuable service to the community and a validation of our efforts here at the Department.

Here are the results of last week's inspections, as well as a week-by-week statistical summary of all inspections during January.

UNSATISFACTORY

The following establishments were found to be in substantial violation of the sanitary code.

Big Daddy's Steak House
431 Grand Avenue, Conover Falls
Inspected January 25, 1999

Toxic chemicals (antifreeze, can of ant/roach killer) found on premises. Potentially hazardous foods not kept at or above 140 degrees F during hot holding. Food not protected—buckets of food stored on floor in cooler, food not

FIGURE 4.8 **Letter Report, page 1**

2

covered in coolers. Raw meat stored over prepared foods in cooler. Food build-up in storage room refrigerator. Canned goods in poor condition (dented, rusted). Bowl used as flour scoop. Box of paper towels improperly stored on floor. Non-food contact surfaces not easily cleanable. Cardboard used as liner on food storage shelves. Restroom missing hand wash sign. Light fixture missing shield and end caps. Kitchen ceiling tiles missing. No 1999 permit on display.

Employee Cafeteria, Paragon Insurance Co.
Airport Road, Cedarville
Inspected January 26, 1999

Potentially hazardous foods not kept at or below 45 degrees F during cold holding. Potentially hazardous foods not kept at or above 140 degrees F during hot holding. Single-service napkins stored on kitchen floor.

Roma Pizzeria
38 Crowley Street, Dunkirk
Inspected January 27, 1999

Worker serving pizza slices with bare hands. Potentially hazardous foods not kept at or above 140 degrees F during hot holding. Food not protected—uncovered food in freezer, salt bucket not labeled. Hair improperly restrained—hats, nets/visors required. In-use utensils stored on paper plate. Employee (delivery driver) smoking in kitchen.

SATISFACTORY

The following establishments were found to be in essential compliance with the sanitary code, although some violations were noted.

Imperial Wok
618 Rogers Street, Cooperton
Inspected January 28, 1999

Potentially hazardous foods not kept at or above 140 degrees F during hot holding. Food not protected—jars of juice stored on kitchen floor. Improper use of utensils—scoop stored handle down in flour.

FIGURE 4.8 Letter Report, page 2

3

Cuzzie's Pub
39 Railroad Street, Monroe
Inspected January 29, 1999

Unshielded light fixture in walk-in cooler. No hand soap in restroom.

NO VIOLATIONS

The following establishments were found to be in full compliance with the sanitary code.

Conover Falls Coffee House
17 Village Green East, Conover Falls
Inspected January 25, 1999

Mister Eight Ball
49 Clinton Street, Dunkirk
Inspected January 27, 1999

SUMMARY OF JANUARY 1999 INSPECTIONS

	Unsatisfactory	Satisfactory	No Violations
Jan. 4–8	3	2	2
Jan. 11–15	5	0	1
Jan. 18–22	2	3	2
Jan. 25–29	3	2	2
Totals	13	7	7

Please feel free to call me at your convenience if you have questions regarding these inspections or any other matters relating to the Monroe County Health Department. Unless I hear otherwise, I will continue to provide inspection results on a weekly basis.

Sincerely,

Marjorie Witkowski

Marjorie Witkowski
Supervisor

FIGURE 4.8 Letter Report, page 3

Monroe County Health Department

MEMORANDUM

DATE: February 9, 1999

TO: Janet Butler, Commissioner

FROM: Marjorie Witkowski, Supervisor

SUBJECT: Inspections Report

As you requested, here is a complete report on the results of Richard Vaughan's inspections of food service establishments in the county during the last week of January, along with a week-by-week statistical summary of his inspections during that month.

FIGURE 4.9 Booklet Report, Cover Memo

INSPECTIONS OF FOOD SERVICE ESTABLISHMENTS
IN MONROE COUNTY, WYOMING
JANUARY 25–29, 1999

Report Submitted to

Janet Butler
Commissioner of Public Health

by

Marjorie Witkowski
Supervisor, County Health Department

February 9, 1999

FIGURE 4.9 Booklet Report, Cover Page

INTRODUCTION

In keeping with its mandate to safeguard the public welfare, the Monroe County Health Department conducts ongoing, unannounced inspections of the County's food service establishments to ensure their compliance with state codes, rules, and regulations. This report provides the results of seven such inspections conducted by Senior Inspector Richard Vaughan during the period of January 25–29 of this year, along with a week-by-week statistical summary of Mr. Vaughan's 27 total inspections during January.

UNSATISFACTORY

The following establishments were found to be in substantial violation of the sanitary code.

Big Daddy's Steak House
431 Grand Avenue, Conover Falls
Inspected January 25, 1999

Toxic chemicals (antifreeze, can of ant/roach killer) found on premises. Potentially hazardous foods not kept at or above 140 degrees F during hot holding. Food not protected—buckets of food stored on floor in cooler, food not covered in coolers. Raw meat stored over prepared foods in cooler. Food build-up in storage room refrigerator. Canned goods in poor condition (dented, rusted). Bowl used as flour scoop. Box of paper towels improperly stored on floor. Non-food contact surfaces not easily cleanable. Cardboard used as liner on food storage shelves. Restroom missing hand wash sign. Light fixture missing shield and end caps. Kitchen ceiling tiles missing. No 1999 permit on display.

Employee Cafeteria, Paragon Insurance Co.
Airport Road, Cedarville
Inspected January 26, 1999

Potentially hazardous foods not kept at or below 45 degrees F during cold holding. Potentially hazardous foods not kept at or above 140 degrees F during hot holding. Single-service napkins stored on kitchen floor.

FIGURE 4.9 **Booklet Report, page 1**

2

Roma Pizzeria
38 Crowley Street, Dunkirk
Inspected January 27, 1999

Worker serving pizza slices with bare hands. Potentially hazardous foods not kept at or above 140 degrees F during hot holding. Food not protected—uncovered food in freezer, salt bucket not labeled. Hair improperly restrained—hats, nets/visors required. In-use utensils stored on paper plate. Employee (delivery driver) smoking in kitchen.

SATISFACTORY

The following establishments were found to be in essential compliance with the sanitary code, although some violations were noted.

Imperial Wok
618 Rogers Street, Cooperton
Inspected January 28, 1999

Potentially hazardous foods not kept at or above 140 degrees F during hot holding. Food not protected—jars of juice stored on kitchen floor. Improper use of utensils—scoop stored handle down in flour.

Cuzzie's Pub
39 Railroad Street, Monroe
Inspected January 29, 1999

Unshielded light fixture in walk-in cooler. No hand soap in restroom.

NO VIOLATIONS

The following establishments were found to be in full compliance with the sanitary code.

Conover Falls Coffee House
17 Village Green East, Conover Falls
Inspected January 25, 1999

FIGURE 4.9 Booklet Report, page 2

3

Mister Eight Ball
49 Clinton Street, Dunkirk
Inspected January 27, 1999

SUMMARY OF JANUARY 1999 INSPECTIONS

	Unsatisfactory	Satisfactory	No Violations
Jan. 4–8	3	2	2
Jan. 11–15	5	0	1
Jan. 18–22	2	3	2
Jan. 25–29	3	2	2
Totals	13	7	7

FIGURE 4.9 Booklet Report, page 3

Types Of Reports

Like memos and letters, workplace reports are written in all kinds of situations, for an enormous variety of reasons. Many reports are in a sense unique because they are written in response to one-time occurrences. On the other hand, it is not uncommon for a given report to be part of an ongoing series of weekly, monthly, or annual reports on the same subject. Most reports can be classified into several broad categories, and some of the most common types of short reports are as follows:

- ***Incident Report.*** Explains the circumstances surrounding a troublesome occurrence such as an accident, fire, equipment malfunction, or security breach.
- ***Progress Report.*** Outlines the status of an ongoing project or undertaking.
- ***Recommendation Report.*** Urges that certain procedures be adopted (or rejected).
- ***Travel Report.*** Identifies the purpose and results of business-related travel.

Of course, an individual report can serve more than one purpose; overlap is not uncommon. An incident report, for example, may well conclude with a recommendations section intended to minimize the likelihood of occurrence. In every situation, the writer must consider the purpose and intended audience for the report. Content, language, tone, degree of detail, and overall approach must be appropriate to the circumstances, and the report headings, formatting, visuals, and other features must suit the role of the particular report. The following pages discuss the four common report types in detail.

Incident Report

An incident report creates a written record of a troublesome occurrence. The report is written either by the person involved in the incident or by the person in charge of the area where it took place. Such a report may be needed to satisfy government regulations, to guard against legal liability, and to draw attention to unsafe or otherwise unsatisfactory conditions in need of correction. Accordingly, an incident report must provide a thorough description of the occurrence, and—if possible—an explanation of the cause(s). In addition, it often includes a section of recommendations for corrective measures to prevent recurrence.

When describing the incident, always provide the complete details:

- names and job titles of all persons involved, including onlookers;
- step-by-step narrative description of the incident;
- exact location of the incident;
- date and exact time of each major development;
- clear identification of any equipment or machinery involved;
- detailed description of any medical intervention required, including names of ambulance services and personnel, nurses, physicians, hospitals, or clinics;
- reliable statements (quotation or paraphrase) from persons involved;
- outcome of the incident.

To avoid liability when discussing possible causes, use qualifiers such as "perhaps," "maybe," "possibly," "it appears." Do not report the comments of witnesses and those involved as if those observations were verified facts; often they are grossly inaccurate. Attribute all such comments to their source(s), and identify them as speculation only. Furthermore, exclude any comments unrelated to the immediate incident. Although you are ethically required to be as complete and accurate as possible, do not create an unnecessarily suspicious climate by relying on second-hand accounts or reporting verbatim the remarks of persons who are obviously angry or distraught, as in this example:

> Ronald Perkins suffered a severed index finger when his left hand became caught in a drill press after he tripped on some wood that another employee had carelessly left on the floor near the machine. According to Perkins, this was "pretty typical of how things are always done around here."

A more objective phrasing might look something like this:

> Ronald Perkins suffered a severed index finger when his left hand became caught in a drill press. Perkins said he had tripped on wood that was lying on the floor near the machine.

Similarly, the recommendations section of an incident report should not seek to assign blame or highlight incompetence, but to encourage the adoption of measures that will decrease the likelihood of repeated problems. Consider, for example, the incident report in Figure 4.10, prepared in memo report format.

Southeast Insurance Company

MEMORANDUM

DATE: October 9, 1998

TO: Jonathan Purdy
Physical Plant Supervisor

FROM: Bonnie Cardillo
Nurse

SUBJECT: Incident Report

John Fitzsimmons, a claims adjuster, slipped and fell in the front lobby of the building, striking his head and momentarily losing consciousness.

DESCRIPTION OF INCIDENT

At approximately 2:55 p.m. on Thursday, October 8, Fitzsimmons was returning from his break when he slipped and fell in the front lobby, striking his head on the stone floor and momentarily losing consciousness. According to Beverly Barrett, the receptionist, the floor had just been mopped and was still wet. She paged Mike Moore, the security officer, who in turn paged me. When I arrived at approximately 3 p.m., Fitzsimmons had revived. I immediately checked his vital signs, which were normal. He refused further medical attention and returned to work. I advised him to contact me if he experienced any subsequent discomfort, but to my knowledge there has been none.

FIGURE 4.10 Incident Report (Memo-Report Format), page 1

2

RECOMMENDATIONS

Two ideas come to mind.

Perhaps we should remind all employees to contact me FIRST (rather than Security) in situations involving personal injury. The sooner I'm contacted, the sooner I can respond. Obviously, time can be an important factor if the problem is serious.

To prevent other occurrences of this nature, perhaps the maintenance staff should be provided with large, brightly colored warning signs alerting employees and public alike to the presence of wet floors. I see these signs in use at the mall, the hospital, and elsewhere, and they do not appear expensive. I have noted also that many are bilingual, bearing both the English warning "Caution: Wet Floor" and the Spanish "Cuidado: Piso Mojado." No doubt they can be ordered from any of the catalogs regularly received by your office.

FIGURE 4.10 **Incident Report (Memo-Report Format) page 2**

Progress Report

A Progress Report provides information about the status of an ongoing project or activity that must be monitored to ensure successful completion within a specified time period. Sometimes called status reports or periodic reports, progress reports are submitted either upon completion of key stages of a project or at regular, preestablished intervals—quarterly, monthly, weekly, or sometimes as often as every day. They are written by the individual(s) directly responsible for the success of the undertaking. The readers of these reports are usually in the management sector of the organization, however, and may not be familiar with the technical details of the situation. Rather, their priority is successful completion of the project within established cost guidelines. Therefore, the information in a progress report tends to be more general than specific, and the language tends to be far less technical than that of other kinds of reports.

Most progress reports include the following components:

- ***Introduction.*** Provides context and background, identifying the project, reviewing its objectives, and alerting the reader to any new developments since the previous progress report.
- ***Work Completed.*** Summarizes accomplishments to date. This section can be organized in either of two ways: if the report deals with one major task, a chronological approach is advisable; if it deals with several related projects, then subdivisions by task are better.
- ***Work Remaining.*** Summarizes all uncompleted tasks, emphasizing what is expected to be accomplished first.
- ***Problems.*** Identifies any delays, cost over-runs, or other unanticipated difficulties. If all is well, or if the problems are of no particular consequence, this section may be omitted.
- ***Conclusion.*** Summarizes the status of the project and recommends solutions to any major problems.

If properly prepared and submitted in a timely fashion, progress reports can be invaluable in enabling management to make necessary adjustments to meet deadlines, avert crises, and prevent unnecessary expense. Figure 4.11 depicts a progress report on capital projects, prepared in booklet report format with a transmittal memo.

WESTON INDUSTRIES, INC.

MEMORANDUM

DATE: November 10, 1999

TO: Robert Tomlinson
Accounting Department

FROM: John Stevens
Physical Plant

SUBJECT: Progress Report on Capital Projects

As requested, here is the progress report on the five capital projects identified as high-priority items at last spring's long-range planning meeting:

- Replacement of front elevator in Main Building
- Replacement of all windows in Main Building
- Installation of new fire alarm system in all buildings
- Installation of emergency lighting system in all buildings
- Renovation of "B" Building basement

Please contact me if you have any questions.

FIGURE 4.11 Progress Report, Booklet Format, Transmittal Memo

WESTON INDUSTRIES, INC.

PROGRESS REPORT
on
CAPITAL PROJECTS

by

John Stevens
Physical Plant

Submitted to

Robert Tomlinson
Accounting Department

November 10, 1999

FIGURE 4.11 **Progress Report, Booklet Format, Title Page**

INTRODUCTION

Weston Industries, Inc. is currently involved in several major capital projects that were identified as high-priority items at last spring's long-range planning meeting: replacement of the front elevator and all windows in the Main Building, installation of a new fire alarm system and emergency lighting system in all buildings, and renovation of the "B" Building basement. Progress has been made on all of these projects, although there have been a few problems.

WORK COMPLETED

Elevator Replacement
Equipment has been ordered from Uptown Elevator. The pump has arrived and is in storage. We have asked Uptown for a construction schedule.

Window Replacement
Entrance and Window Wall: KlearVue Window Co. has completed this job, but it is unsatisfactory. See "Problems" section, below. Other Windows: Architect has approved submittal package and Cavan Glass Co. is preparing shop drawings. Architect has sent Cavan Glass Co. a letter stating that work must begin no later than April 1, with completion in July.

Fire Alarm System
First submittal package from Alert-All, Ltd. was reviewed by architect and rejected. A second package was accepted. The alarm system is on order.

Emergency Lighting System
BriteLite, Inc. has begun installation in the Main Building. The plan is to proceed on a building-by-building basis, completing one before moving on to another.

Basement Renovation
First submittal package from Innovation Renovation was reviewed by architect and rejected. Innovation Renovation is preparing a second package in an attempt to reduce the cost of HVAC work. Work is expected to begin in June.

FIGURE 4.11 Progress Report, Booklet Format, page 1

2

WORK REMAINING

Elevator Replacement
Construction schedule must be received from Uptown Elevator. Work must begin.

Window Replacement
Entrance and Window Wall: Problems with KlearVue Window Co. must be resolved. See "Problems" section, below.

Other Windows: Shop drawings must be received from Cavan Glass Co. and approved. Work must begin.

Fire Alarm System System must be received. Work must begin. Work will be completed during down time (10 p.m. to 6 a.m.) to minimize disruption.

Emergency Lighting System
BriteLite, Inc. must complete installation in Main Building, then move on to other buildings. Like the fire alarm installation, the bulk of this work will be done during down time.

Basement Renovation
Final submittal package must be received from Innovation Renovation and approved. Work must begin.

PROBLEMS

Window Replacement
Entrance and Window Wall: KlearVue Window Co. is still responsible for replacing one window that has a defect in the glass. In addition, the architect refuses to accept three of the five large panes in the window wall due to excessive distortion in the glass. The architect has sent several letters to KlearVue but has received no response. The remaining balance on this contract ($15,750) is therefore being held, pending resolution of these problems.

CONCLUSION

Although none of the five capital projects targeted at the spring meeting has in fact been satisfactorily completed, all but one are moving forward through expected channels. The one troublesome item—the unsatisfactory windows—should be resolved. If KlearVue continues to ignore the architect's inquiries, perhaps our attorneys should attempt to get a response.

FIGURE 4.11 Progress Report, Booklet Format, page 2

Recommendation Report

A recommendation report assesses a troublesome or unsatisfactory situation, identifies a solution to the problem(s), and persuades decision-makers to pursue a particular course of action that will improve matters. Such reports are sometimes unsolicited. Generally, however, a recommendation report is written by a knowledgeable employee who has been specifically assigned the task. As with most kinds of reports, the content can vary greatly depending on the nature of the business or organization, and also upon the nature of the situation at hand. In nearly all cases, however, recommendation reports are intended to enhance the quality of products or services, to maximize profits, to reduce costs, or to improve working conditions.

In the case of a solicited report, the writer should attempt to get a written request from the individual who wants the report, then carefully study it to determine the exact parameters of the situation in question. If unsure of any aspect of the assignment, the writer should seek clarification before continuing. As discussed in Chapter 1, it is vital to establish a firm sense of purpose and audience before you attempt to compose any workplace writing. A clear and focused written request—or the discussion generated by the lack of one—will provide guidance in this regard.

Since recommendation reports are persuasive in nature, they are in several respects trickier to write—and live with afterwards—than reports that are intended primarily to record factual information. Tact is of great importance. Since your report essentially is designed to bring about an improvement in existing conditions or procedures, you should guard against appearing overly critical of the present circumstances. Focus more on what *will be* than on what *is*. Emphasize solutions rather than problems. Do not assign blame for present difficulties except in the most extreme cases. A very helpful strategy in writing recommendation reports is to request input from co-workers, whose perspective may give you a more comprehensive understanding of the situation you are assessing.

Recommendation reports are structured in various ways, but almost all include three basic components.

- ***Problem.*** It is important to identify not only the problem itself, but also—if possible—its causes and its relative urgency.
- ***Solution.*** This section not only sets forth a recommendation, but also explains how it will be implemented and clearly states its advantages, including relevant data on costs, timing, and the like.

- *Discussion.* This brief overview summarizes the report's key points and politely urges the adoption of its recommendation.

Figure 4.12, prepared in letter report format, depicts a recommendation report designed to enable a feed manufacturing company to avert fiscal problems by cutting costs at one of its mills.

Travel Report

There are two kinds of travel reports: field reports and trip reports. The purpose of both is to create a record of—and, by implication, justification for—an employee's work-related travel. The travel may be directly related to the performance of routine duties (a field visit to a customer or client, for example) or it may be part of the employee's ongoing professional development, such as a trip to a convention, trade show, or off-site training session. Submitted to the employee's immediate supervisor, such a report not only describes the employee activity made possible by traveling, but also assesses the activity's value and relevance to the organization.

Travel reports are usually structured as follows.

- *Introduction.* Provides all basic information, including destination, purpose of travel, arrival and departure dates and time, and mode of travel (personal car, company car, train, plane)
- *Description of Activity/Service Performed.* Not an itinerary, but rather a selectively detailed account. The degree of detail is greater if readers other than the supervisor will have access to the report and expect to learn something from it. In the case of a field report, any problems encountered should be detailed, along with corrective actions taken.
- *Cost Accounting.* In the case of non-routine travel, the employee is usually expected to account for all money spent, especially if it is to be reimbursed by the employer.
- *Discussion.* Assesses the usefulness of the travel, and—if applicable—makes recommendations regarding the feasibility of other such travel in the future. In the case of a field report, suggestions are sometimes made based on the particulars of the situation.

Figures 4.13 and 4.14, both in memo report format, depict the two kinds of travel reports.

COOPER & SONS FEED COMPANY

"Serving Livestock Breeders Since 1932"

Des Moines Mill • State Highway, Des Moines, IA 50300 • (515) 555-1234

February 9, 1999

Ms. Mary Cooper, C.E.O.
Cooper & Sons Feed Company
Main Office
427 Cosgrove Street
Des Moines, IA 50300

Dear Ms. Cooper:

Here is the report you requested, outlining a proposed expense management plan that will enable the Des Moines Mill to cut costs.

PROBLEM

Because of the recent closings of several large family-run farms in the surrounding area, our profit margin has shrunk. We must therefore reduce the Des Moines Mill's annual operating budget by at least $70,000 for it to remain viable.

SOLUTION

Inventory Reduction
Reduce inventory by $50,000 thereby creating savings on 10% interest expense.

Saving: $5,000.

Elimination of Hourly Position
Based on seniority, eliminate one Customer Service position, distributing responsibilities between the two remaining.

Saving: $10,500 in wages plus $2,100 benefits; total $12,600

FIGURE 4.12 **Recommendation Report (Letter Report Format), page 1**

2

Elimination of Salaried Position
Eliminate Plant Manager's position, distributing responsibilities between the two Assistant Managers.

Saving: $36,320 salary plus $7,264 benefits; total $43,584

Reduction of Remill Costs
Each load returned from farm for remill costs an average of $165 and creates 3.5 hours of overtime work. Lowering our error rate from 2 per month to 1 per month will save $1,980 annually. In addition, this will raise the ingredient value we capture on these feeds by 50% (6 ton/month x 12 months x $100 increased value) or $7,200.

Saving: total $9,200

DISCUSSION

Adoption of the above measures will result in a total annual savings of $70,384. This is within the requirements.

The principal negative impact will be on personnel, and we regret the necessity of eliminating the two positions. It should be noted, however, that the situation could be much worse. The hourly Customer Service employee can be re-hired after the scheduled retirement of another Customer Service worker next year. Also, the retrenched Plant Manager can be offered a comparable position at the Cooper & Sons mill in Northton, where business is booming and several openings currently exist.

Therefore, the above measures should be implemented as soon as possible, to ensure the continued cost-effectiveness of the Des Moines mill.

Thank you for considering these recommendations. I appreciate having the opportunity to provide input that may be helpful in the Company's decision-making process.

Sincerely,

John Svenson

John Svenson
Operations Assistant

FIGURE 4.12 Recommendation Report (Letter Report Format), page 2

ACE TECHNOLOGIES CORP.

MEMORANDUM

DATE: November 12, 1999

TO: Joseph Chen, Director
Sales & Service

FROM: Peter Hutchins
Service Technician

SUBJECT: Travel to Jane's Homestyle Restaurant (Account #2468)

INTRODUCTION

On Monday, November 9, I traveled by company truck to Jane's Homestyle Restaurant in Northweston to investigate the owner's complaint regarding malfunctioning video monitors (Ace Cash Register System 2000). I left the plant at 9 a.m. and was back by 10:30.

SERVICE PERFORMED

All three video monitors were functioning erratically. Upon examination, however, the problem turned out to be very simple. Because of how the Jane's Homestyle Restaurant counter area is designed, the keypad must be positioned farther away from the monitor than usual. As a result, the 15" cable (part #012) that creates interface between the two units is not quite long enough to stay firmly in place, making the connection unstable.

After explaining the problem to the restaurant manager, I provided a temporary "quick fix" by duct-taping the connections. Upon returning to the plant I instructed the shipping department to send the restaurant three 20" replacement cables (part #123) by overnight delivery.

DISCUSSION

This incident demonstrates the need for more thorough testing of our systems when they are initially installed, taking fully into account all features of the environments in which they will be used. We might check with the other System 2000 accounts to determine whether any other customers have experienced similar difficulty. Maybe all System 2000 units should routinely be installed with the longer cable.

FIGURE 4.13 Travel Report (Field Report)

ACE TECHNOLOGIES CORP.

MEMORANDUM

DATE: November 12, 1999

TO: Floyd Danvers, Director
Human Resources

FROM: Peter Hutchins
Service Technician

SUBJECT: Travel to Northweston Marriott for Seminar

INTRODUCTION

On Thursday, November 4, and Friday, November 5, I traveled by company car to the Northweston Marriott to attend a seminar entitled "Workplace Communications: The Basics," presented by a corporate training consultant, Dr. George J. Searles. I left the plant at 8 a.m. and was back by 5 p.m. both days.

ACTIVITIES

The seminar consisted of four half-day sessions, as follows:

- Workplace Communications Overview (Thursday a.m.)
- Review of Mechanics (Thursday p.m.)
- Memos and Letters (Friday a.m.)
- Reports (Friday p.m.)

There were 21 participants from a variety of local businesses and organizations, and the sessions were a blend of lecture and discussion, with emphasis on clear, concise writing. The instructor distributed numerous handouts that illustrated the points under consideration.

COSTS

The program cost $500, paid by the company. Aside from two days' lunch allowance ($10 total) and use of the company car (38 miles total), there were no other expenses.

DISCUSSION

This was a very worthwhile program. I learned a lot from it. Since it would be quite difficult, however, to summarize the content here, I've appended a complete set of the handouts distributed by the instructor. As you will see when you examine these materials, the whole focus of the program was quite practical and hands-on. I recommend that other employees be encouraged to attend the next time this program is offered in our area.

FIGURE 4.14 Travel Report (Professional Development Trip)

☑ Checklist Evaluating a Memo Report

A good memo report

___ follows standard memo report format;

___ includes certain features:

- ☐ the word MEMO or MEMORANDUM as a heading or near the top,
- ☐ TO line, which provides the full name, title, and department of the receiver,
- ☐ FROM line, which provides the full name, title, and department of the sender,
- ☐ DATE line,
- ☐ SUBJECT line, which provides a clear, and accurate, but brief indication of what the memo report is about;

___ is organized into separate, labeled sections that cover the subject fully in a well-organized way;

___ includes no inappropriate content;

___ uses clear, simple language;

___ maintains an appropriate tone, neither too formal nor too conversational;

___ employs effective visuals—tables, graphs, charts, etc.—where necessary to clarify the text;

___ contains no typos or mechanical errors in spelling, capitalization, punctuation, and grammar.

Exercises

■ EXERCISE 4.1

Write a report either to your supervisor at work or to the campus safety committee at your college, fully describing the circumstances surrounding an accident or injury you've experienced at work or at college, and the results of that mishap. Include suggestions about how similar situations might be avoided in the future. Use the memo report format and include visuals if appropriate.

☑ Checklist Evaluating a Letter Report

A good letter report

___ follows standard letter format;

___ includes certain features:

- ☐ sender's complete address,
- ☐ date,
- ☐ receiver's full name and complete address,
- ☐ salutation, followed by a colon,
- ☐ complimentary close, followed by a comma,
- ☐ sender's signature and full name;

___ is organized into separate, labeled sections, covering the subject fully in an orderly way:

- ☐ first paragraph establishes context and states the purpose,
- ☐ middle paragraphs provide the report, separated into labeled sections that cover the subject matter fully in a well-organized way, and incorporate any necessary visuals,
- ☐ last paragraph politely achieves closure;

___ includes no inappropriate content;

___ uses clear, simple language;

___ maintains an appropriate tone, neither too formal nor too conversational;

___ employs effective visuals—tables, graphs, charts, etc.—where necessary to clarify the text;

___ contains no typos or mechanical errors in spelling, capitalization, punctuation, and grammar.

■ EXERCISE 4.2

Write a report to the local police department, studying the rush-hour traffic patterns at a major intersection near campus. Observe for one hour during either the morning or evening rush period on one typical weekday. Record the number and kinds of vehicles (car, truck, bus, motorcycle) and the directions in which they were traveling, along with an estimate of pedestrian traffic. Also record, of course,

☑ Checklist Evaluating a Booklet Report

A good booklet report

___ is accompanied by a transmittal document (memo or letter);

___ includes a title page that contains the following:

- ☐ title of the report,
- ☐ name(s) of author(s),
- ☐ name of company or organization,
- ☐ name(s) of person(s) receiving the report,
- ☐ date;

___ is organized into separate, labeled sections that cover the subject fully in a well-organized way;

___ includes no inappropriate content;

___ uses clear, simple language;

___ maintains an appropriate tone, neither too formal nor too conversational;

___ employs effective visuals—tables, graphs, charts, etc.—where necessary to clarify the text;

___ contains no typos or mechanical errors in spelling, capitalization, punctuation, and grammar.

any accidents that occur. Evaluate the layout of the intersection (including lights, signs, and so forth) in terms of safety, and suggest improvements. Use the booklet report format and include visuals.

■ EXERCISE 4.3

Write a report to your communications instructor, outlining your progress in class. List attendance, grades, and any other pertinent information, including an objective assessment of your performance so far, and the final grade you anticipate receiving. Use the memo report format and include visuals.

■ EXERCISE 4.4

Write a report to the Academic Dean, urging that a particular college policy be modified. Be specific about the reasons for your proposal. Justify the change and provide concrete suggestions about possible alternative policies. Use the memo report format and include visuals if appropriate.

■ EXERCISE 4.5

Write a report to your instructor, discussing any recent vacation trip you have taken. Summarize your principal activities during the trip, and provide an evaluation of how successful the vacation was. Use the letter report format and include visuals.

■ EXERCISE 4.6

Write a report to a classmate, outlining the performance of your favorite sports team over the past three years. Using statistical data, be as factual and detailed as your knowledge of the sport will permit. Attempt to explain the reasons for the team's relative success or lack of it. Use the booklet report format and include visuals.

■ EXERCISE 4.7

Write a report to the Student Services Director or the Physical Plant Director at your college, evaluating a major campus building with respect to accessibility to the physically challenged. Discuss the presence or absence of special signs, doors, ramps, elevators, restroom facilities, and the like. Suggest additional accommodations that should be provided if such needs exist. Use the booklet report format and include visuals.

■ EXERCISE 4.8

Team up with a classmate of the opposite sex, and write a report to the Physical Plant Director analyzing the differences—if any—between the men's and women's restroom facilities in the main building on your campus. Suggest any changes or improvements you think might be necessary. Use the booklet report format and include visuals.

■ EXERCISE 4.9

Have you ever been the victim of or witness to a minor crime on campus? Write a report to the College Security Director, relating the details of that experience, and offering suggestions about how to minimize the likelihood of similar occurrences in the future. Use the letter report format and include visuals if appropriate.

■ EXERCISE 4.10

Write a report to your classmates in which you evaluate three nearby restaurants featuring similar cuisine (for example, seafood, Chinese, Italian) or three nearby stores that sell essentially the same product (for example, athletic shoes, books and music, clothing). Discuss such issues as selection, quality, price, and service. Use the booklet report format and include visuals.

■ EXERCISE 4.11

Each of exercises 4.1–4.10 could be described as belonging to one (or more) of the report categories discussed in this chapter: Incident, Progress, Recommendation, and Travel. Write your instructor a report in which you categorize each of the exercises, providing reasons for your choices. Use the memo report format.

■ EXERCISE 4.12

The writer of the fictitious letter report depicted on the following pages has over-used the design options afforded by the computer, creating an extremely confusing report that's almost impossible to comprehend. Fix it.

■ **EXERCISE 4.12** **Continued**

Social Security Administration
Supplemental Security Income
Important Information

Date: November 10, 1999

Claim Number: 123-45-6789 DI

COUNTY DSS FOR
PETER WOOD
NORTHTON MN 55100

Type of Payment
Individual—Disabled

We are writing to tell you about changes in Peter Wood's Supplemental Security Income record.

INFORMATION ABOUT Peter Wood's PAYMENTS

This information does not change his current payment amount.

PETER WOOD'S Payment Is Based on These Facts

He has monthly income which must be considered in figuring his eligibility as follows:

••• His other unearned income of $157.30 for April 1999 on.
*** His wages of $75.00 or less for April 1999 on.

Things to Remember

This information is also being sent to Peter Wood.

EXERCISE 4.12 Continued

Do you disagree with the Decision???

If you disagree with the decision, you have the right to appeal. We will review your case and consider any new facts you have.

- You have 60 days to ask for an appeal.
- The 60 day period starts the day AFTER you get this letter.
- You must have GOOD REASON for waiting more than 60 days.

TO APPEAL YOU MUST FILL OUT A FORM CALLED "REQUEST FOR RECONSIDERATION." THE FORM NUMBER IS SSA-561. TO GET THIS FORM CONTACT ONE OF OUR OFFICES. WE CAN HELP YOU FILL OUT THE FORM.

HOW TO APPEAL

There are two ways to appeal. You can pick the one you want. If you meet with us in person, it may help us decide your case.

* CASE REVIEW: *You have a right to review the facts in your file. You can give us more facts to add to your file. Then we'll decide your case again. You won't meet with the person who decides your case. This is the only kind of appeal you can have to appeal a medical decision.*

* INFORMAL CONFERENCE: *You'll meet with the person who decides your case. You can tell that person why you think you're right. You can give us more facts to help prove you're right. You can bring other people to help explain your case.*

If you want help with your appeal

YOU CAN HAVE A FRIEND, LAWYER, OR SOMEONE ELSE HELP YOU. THERE ARE GROUPS THAT CAN HELP YOU FIND A LAWYER OR GIVE YOU FREE LEGAL SERVICES IF YOU QUALIFY. THERE ARE ALSO LAWYERS WHO DO NOT CHARGE UNLESS YOU WIN YOUR APPEAL. YOUR LOCAL SOCIAL SECURITY OFFICE HAS A LIST OF GROUPS THAT CAN HELP WITH YOUR APPEAL. IF YOU GET SOMEONE TO HELP YOU, YOU SHOULD LET US KNOW. IF YOU HIRE SOMEONE, WE MUST APPROVE THE FEE.

EXERCISE 4.12 Continued

IF YOU HAVE ANY QUESTIONS

If you have any questions, you may call us toll-free at 1-800-555-1234, or call your local Social Security Office at 1-612-555-1234. We can answer most questions over the phone. You can also write or visit any Social Security office. The office that serves your area is located at:

BRANCH OFFICE
123 NORTHTON AVE
NORTHTON, MN 55100

If you do call or visit an office, please have this letter with you. It will help us answer your questions. Also, if you plan to visit an office, you may call ahead to make an appointment. This will help us serve you more quickly when you arrive at the office.

Sincerely,

Neil Perrault

Neil Perrault

Supplemental Security Administrator

5

Summaries

Learning Objective Upon completing this chapter, you will be able to write clear, concise, and complete summaries that convey the content and emphasis of the original sources.

In the broadest sense, *all* writing is a form of summary. Whenever we put words on paper or computer screen, we condense ideas and information to make them coherent to the reader. Ordinarily, however, the term "summary" refers to a brief statement of the essential content of something heard, seen, or read. For any kind of summary the writer reduces a body of material to its bare essentials. Creating a summary is therefore an exercise in *compression,* requiring logical organization, clear and concrete terminology, and sensitivity to the reader's needs—as does any other kind of workplace communication. Summary writing, however, demands an especially keen sense not only of what to include but also of what to *leave out.* The goal is to highlight the key points and not burden the reader with unnecessary details. In the workplace context, the most common summary application is in the abstracts and executive summaries that accompany long reports. This chapter will explore the main principles governing the writing of summaries, a valuable skill in many work settings.

Types of Summaries: Descriptive, Informative, and Evaluative

In general, summaries can be classified into three categories: descriptive, informative, and evaluative.

A **descriptive summary** states what the original document is about, but does not convey any of the specific information contained therein. It is much like a table of contents in paragraph form. Its main purpose is to help a reader to determine whether the document summarized is of any potential use in a given situation. For example, a Government Printing Office pamphlet providing descriptive summaries of publications about workplace safety may be quite helpful to a personnel director wishing to educate employees about a particular job-related hazard. Similarly, a purchasing agent may consult descriptive summaries to determine the potential relevance of outside studies of needed equipment or supplies. A descriptive summary may look something like this:

> This report discusses a series of tests conducted on industrial-strength coil springs at the TopTech Laboratories in Northton, Minnesota, in January 1999. Three kinds of springs were evaluated for flexibility, durability, and heat resistance, to determine their relative suitability for several specific manufacturing applications at Northton Industries.

After reading this summary, someone seeking to become better informed about the broad topic of coil springs may decide to read the report.

An **informative summary**, on the other hand, goes considerably further and presents the document's content, although in greatly compressed form. A good informative summary that includes the document's conclusions and recommendations (if any) can actually enable a busy reader to *skip* the original altogether. Here is an informative version of the descriptive summary shown above:

> This report discusses a series of tests conducted on industrial-strength coil springs at the TopTech Laboratories in Northton, Minnesota, in January 1999. Three kinds of springs—all manufactured by the Mathers Spring Co. of Marietta, Ohio—were tested: serial numbers 423, 424, and 425. The springs were evaluated for flexibility, durability, and heat resistance, to determine their relative suitability for several specific manufacturing applications at Northton Industries. In 15 tests using a Flexor Meter, #423 was found to be the most flexible, followed by #425 and #424 respectively. In 15 tests using a Duro Meter, #425 proved the most durable, followed by #423 and #424 respectively. In 15 tests using a Thermal Chamber, #423 was the most heat-resistant, followed by #424 and #425 respectively. Although #423 compiled the best overall performance rating, #425 is the preferred choice, as the applications in question require considerable durability and involve relatively few high-temperature operations.

The **evaluative summary** is even more fully developed and includes the writer's personal assessment of the original document. The following is an evaluative version of the same summary. Notice that the writer inserts subjective value judgments throughout.

> This rather poorly written and finally unreliable report discusses a series of flawed experiments conducted upon industrial-strength coil springs at the TopTech Laboratories in Northton, Minnesota, in January 1999. Three kinds of springs—all manufactured by the Mathers Spring Co. of Marietta, Ohio—were tested: serial numbers 423, 424, & 425. The springs were evaluated for flexibility, durability, and heat resistance, to determine their relative suitability for several specific manufacturing applications at Northton Industries. In 15 tests using the notoriously unreliable Flexor Meter, #423 was rated the most flexible, followed by

> #425 and #424 respectively. In 15 tests using the equally outdated Duro Meter, #425 scored highest, followed by #423 and #424 respectively. In 15 tests using a state-of-the-art Thermal Chamber, #423 was found to be the most heat-resistant, followed by #424 and #425 respectively. Although #423 compiled the best overall performance rating, the report recommends #425, on the grounds that the specific applications in question require considerable durability and involve relatively few high-temperature operations. But these conclusions are questionable at best. TopTech Laboratories has since been closed after revelations of improper procedure, two of the three test sequences involved obsolete instruments, and #425 proved markedly inferior to both #423 and #424 in the only one of the three test sequences that can be considered reliable.

Of the three categories, the informative summary is by far the most common. As in a *Reader's Digest* condensed version of a longer original, its purpose is to convey the main ideas of the original in shorter form. To make an informative summary concrete and to-the-point rather than vague and rambling, be sure to include hard data—such as names, dates, and statistics—as well as the original document's conclusions and recommendations, if any. Sometimes including a good, well-focused quotation from the original can also be very helpful to the reader. Avoid lengthy examples and sidetracks, however, as a summary must always be *brief,* usually no more than a quarter of the original document's length.

In addition, a summary should retain the *emphasis* of the original. For example, a relatively minor point in the source should not take on disproportionate significance in the summary (and perhaps should be omitted altogether). However, crucial information in the original should be equally prominent in the summary. And all information in the summary should spring directly from something in the source. No new or additional information should appear, nor should personal opinion or comments be included unless the summary's purpose is to evaluate.

For the sake of clarity, all workplace writing should be worded in the simplest possible terms. This is especially important in a summary, which is meant to stand alone. If the reader must go back to the original to understand, then the summary is a failure. Hence the summary should be coherently organized and should be written in complete sentences whose meaning is unmistakably clear.

Depending on its nature, a summary that accompanies a long report is called an "abstract" or "executive summary." If the summary is intended simply to provide a general overview of the report, it appears near the beginning of the report and is called an abstract. For an example of such a summary, see Figure 10.1, page 242. If the summary is intended to assist management in decision-making without the managers necessarily having read the report it precedes, it is called an executive summary.

Summarizing Print Sources

To summarize information that already exists in written form, follow these simple steps:

1. Read the entire piece straight through to get a general sense of its content. Pay particular attention to the introduction and the conclusion.
2. Watch for "context clues" (title, subheadings, visuals, boldface print, etc.) to ensure an accurate understanding of the piece.
3. Go back and underline or highlight the most important sentence(s) in each paragraph. Write down all those sentences.
4. Now edit the sentences you selected, compressing, combining, and streamlining wherever you can. You can condense sentences by simplifying them in your own words, or abridge them, largely retaining the original wording.
5. Reread your summary to see if it flows smoothly. Insert transitions—such as "therefore," "however," "nevertheless"—where necessary to eliminate any abrupt jumps from idea to idea.
6. Include concrete facts such as names, dates, statistics, conclusions, and recommendations. (This is especially important because a summary is typically written as one lengthy paragraph incorporating many ideas.)
7. Correct all typos and mechanical errors in spelling, capitalization, punctuation, and grammar.

Figures 5.1–5.4 depict the major steps in the creation of an effective summary from an existing text, in this case a one-page magazine article about the evolution of the modern voting machine.

Since the earliest days of democracy, as the world has sought the ideal form of government, it has also sought the ideal method of voting. The ancient Greeks decided public questions by clashing spears on shields. Colonial America at first favored the show of hands, or splitting into groups. Later the *viva voce* method, in which a voter would openly declare his preference (and then be thanked in florid fashion by the candidate), gained favor. New England pioneered the secret election, sometimes using grains of corn or beans to signify a yes or no and sometimes using written ballots. Down South, where fewer people could read, open voting persisted in some states until after the Civil War, when literacy requirements became politically useful.

By the mid-1800s paper ballots were widespread, and along with the advantage of secrecy came the greater possibility of fraud. Corrupt political operators stuffed ballot boxes and destroyed opposing votes. To combat such tactics, various inventors turned their attention to mechanical voting. As early as 1849 Jan Josef Baranowski of France described his *Scrufateur Mécanique,* and a decade later Werner von Siemens of Germany built a primitive *Abstimmungsapparate.*

Around 1870 at least four different models were used briefly in Great Britain, and more than a hundred American patents were issued for voting machines in the 1860s and 1870s. In most of these the voter pushed a button or inserted a key to drop a ball into a designated bin. Such devices promised to eliminate spoiled and ambiguous ballots, but the balls still had to be tallied, and the possibility of intentional or inadvertent miscounts remained. Few communities judged the machines worth the expense.

The modern voting machine was introduced in a local election in

An 1894 voter poses with the new machine.

Lockport, New York, on April 12, 1892. It was designed by Jacob H. Myers of Rochester, a maker of theft-proof bank safes, who saw his new invention as performing a similar function: fighting vote thieves. Its festive inauguration attracted the sort of turnout normally seen only for a presidential race, including "many aged men, also crippled men," according to a report by the town board. The New York *World* and Rochester *Herald* sent reporters.

Myers's wood-and-steel contraption was ten feet square, illuminated inside with an oil lamp. Except for size it was quite similar to today's machines. A voter entered, locked the door behind him, selected candidates from the Democratic, Republican, or Prohibition party by punching keys, and exited through another door, recording his choice by slamming it firmly. Votes were automatically totaled on numerical registers. No tedious counting was necessary, so the possibility of error or fraud was virtually eliminated. There were sixty candidates and two questions on the ballot, but the complete results were announced ten minutes after the polls closed.

The innovation spread rapidly through upstate New York. Rochester used more than sixty-five voting machines in 1896, and by 1904 twenty cities and many more towns had made the switch. In 1920 more than half the state's population outside New York City voted by machine.

Progress elsewhere was less rapid. The prospect of honest elections did not always appeal to politicians, and the price of the machines—$600 at the turn of the century, $750 to $1,000 in the 1920s—was another obstacle. Proponents could point to the savings in printing, personnel, and litigation costs, but these all afforded ward heelers promising opportunities for dispensing favors and collecting graft.

In addition, there was the usual resistance to anything new. Voters worried that machines would not record their votes properly, though humans did a far less accurate job. As late as 1938 the Kentucky Supreme Court ruled that the state constitution required paper ballots. But as the twentieth century progressed, people became more comfortable with machines of all sorts. In the 1960 presidential election more than half the votes were recorded mechanically.

Today virtually all American elections are conducted by machines, but those of the Myers type are falling out of favor. Punch-card voting was introduced in Ohio in 1960, and optical scanners and video terminals have also recently become popular. As computers perform more and more of the work that used to be done by the mayor's son-in-law, fraud and incompetence are even less of a factor.

Mechanization can affect people's basic rights in various ways, some good and some not. But for a century now, one of the most fundamental rights of all—the right to vote—has been safeguarded by machines. ★

FIGURE 5.1 Article With Most Important Sentences Underlined

Source: From Frederic D. Schwarz. "Machine Politics." *Invention and Technology* 7.9 (Spring 1992), p. 64.

Since the earliest days of democracy, as the world has sought the ideal form of government, it has also sought the ideal method of voting.

By the mid-1800s paper ballots were widespread, and along with the advantage of secrecy came the greater possibility of fraud.

To combat such tactics, various inventors turned their attention to mechanical voting.

Around 1870 at least four different models were used briefly in Great Britain, and more than a hundred American patents were issued for voting machines in the 1860s and 1870s.

Few communities judged the machines worth the expense.

The modern voting machine was introduced in a local election in Lockport, New York, on April 12, 1892.

It was designed by Jacob H. Myers of Rochester, a maker of theftproof bank safes, who saw his new invention as performing a similar function: fighting vote thieves.

Except for size it was quite similar to today's machines.

Votes were automatically totaled on numerical registers.

No tedious counting was necessary, so the possibility of error or fraud was virtually eliminated.

The innovation spread rapidly throughout upstate New York.

Progress elsewhere was less rapid.

The prospect of honest elections did not always appeal to politicians, and the price of the machines—$600 at the turn of the century, $750 to $1,000 in the 1920s—was another obstacle.

In addition, there was the usual resistance to anything new.

But as the twentieth century progressed, people became more comfortable with machines of all sorts.

Today virtually all American elections are conducted by machines, but those of the Myers type are falling out of favor.

Punch-card voting was introduced in Ohio in 1960, and optical scanners and video terminals have also recently become popular.

But for a century now, one of the most fundamental rights of all—the right to vote—has been safeguarded by machines.

FIGURE 5.2 Compilation of Article's Most Important Sentences

Since the dawn of democracy, the world has sought the ideal method of voting.

By 1850 paper ballots were widespread, and this increased the possibility of fraud.

Therefore, inventors turned their attention to mechanical voting.

During the late 1800s, voting machines of various kinds were used in Great Britain and the United States.

But few communities judged these early machines worth the expense.

The modern voting machine was introduced in a Lockport, New York, election on April 12, 1892.

Designed by Jacob H. Myers of Rochester, it was quite similar to today's machines, virtually eliminating the likelihood of error.

The innovation was not popular everywhere; honest elections did not always appeal to politicians, the high price of the machines was another obstacle, and there was the usual resistance to anything new.

As the twentieth century progressed, however, people became more comfortable with machines of all sorts, and now virtually all American elections are conducted mechanically.

In recent years, voting machines equipped with optical scanners and video terminals have become popular.

For a century now, our right to vote has been safeguarded by machines.

FIGURE 5.3 Article's Most Important Sentences, Edited

County Community College

MEMORANDUM

DATE: March 25, 1999

TO: Professor Mary Ann Evans, Ph.D.
English Department

FROM: George Elliot, Student

SUBJECT: Summary

In fulfillment of the "summary" assignment in English 110, Workplace Communication, here is a memo report. I have summarized a magazine article, "Machine Politics," by Frederic D. Schwarz. It was published in the Spring 1992 issue (Vol. 7, No. 4) of *Invention and Technology* on pg. 64. The article is attached.

Machine Politics

Since the dawn of democracy, the world has sought the ideal method of voting. By 1850 paper ballots were widespread, and this increased the possibility of fraud. Therefore, inventors turned their attention to mechanical voting. During the late 1800s, voting machines of various kinds were used in Great Britain and the United States. But few communities judged these early machines worth the expense. The modern voting machine was introduced in a Lockport, New York, election on April 12, 1892. Designed by Jacob H. Myers of Rochester, it was quite similar to today's machines, virtually eliminating the likelihood of error. The innovation was not popular everywhere; honest elections did not always appeal to politicians, the high price of the machines was another obstacle, and there was the usual resistance to anything new. As the twentieth century progressed, however, people became more comfortable with machines of all sorts, and now virtually all American elections are conducted mechanically. In recent years, voting machines equipped with optical scanners and video terminals have become popular. For a century now, our right to vote has been safeguarded by machines.

FIGURE 5.4 Summary (Memo Report Format)

Summarizing Non-Print Sources

To summarize a speech, briefing, broadcast, or other oral presentation for which no transcript exists, you must rely on your own notes. Therefore, you should develop some sort of personal "shorthand" system incorporating abbreviations, symbols, and other notations, to enable you to take notes quickly without missing anything important. Figure 5.5 lists twenty such shortcuts. Probably you will develop others of your own. This strategy is no help, however, if you have to *think* about it. To serve its purpose, it has to become instinctive. Furthermore, you must be able to translate your shortcuts back into regular English as you review your notes. Like anything, this process becomes easier with practice.

FIGURE 5.5 **Note-Taking Shortcuts**

Notation	Meaning	Explanation
=	is	symbol instead of word
#	number	symbol instead of word
&	and	symbol instead of word
∴	therefore	symbol instead of word
2	to, too, two	numeral instead of word
4	for, four	numeral instead of word
B	be, bee	letter instead of word
C	see, sea	letter instead of word
U	you	letter instead of word
Y	why	letter instead of word
R	are	letter instead of word
R̸	are not	slash to express negation
w.	with	abbreviation
w̸.	without	slash to express negation
bcs	because	elimination of vowels
2B	to be	blend of numeral and letter
B4	before	blend of letter and numeral
rathan	rather than	blend of two words
rite	right	phonetic spelling
turn handle	turn the handle	elimination of obvious

When preparing to summarize from non-print sources, some writers tape record the oral presentation and take detailed notes afterwards, at their own pace, under less pressured conditions. This is sometimes a good strategy, especially when time is not a factor or when exact quotation is legally necessary. But even when using a tape recorder, it is still important to take some notes during the actual presentation. A few notations indicating the main points are sufficient, for you can later refer to those sections of the tape to retrieve the particulars. Searching for those sections can be time-consuming, however, unless you use the tape recorder's counter (similar to the mileage odometer on an automobile) and include counter numbers in your notes. If your notes indicate, for example, that Point A was discussed when the counter was at 075, Point B was discussed when the counter was at 190, and Point C was discussed when the counter was at 255, it is much easier to locate the desired sections of the tape.

☑ Checklist Evaluating a Summary

A good summary

___ is no more than 25 percent as long as the original;

___ accurately reports the main points of the original;

___ includes no minor or unnecessary details;

___ includes nothing extraneous to the original;

___ preserves the proportion and emphasis of the original;

___ is well organized, providing transitions to smooth the jumps from idea to idea;

___ maintains an objective tone;

___ uses clear, simple language;

___ contains no typos or mechanical errors in spelling, capitalization, punctuation, and grammar.

Exercises

EXERCISE 5.1

Here are three summaries of the same article, Bruce Porter's "Pete Hamill Wakes Up the Daily News" in the May/June 1997 issue of *The Columbia Journalism Review*. Identify each of the three as descriptive, informative, or evaluative in nature.

Summary A

Veteran New York journalist Pete Hamill implemented major changes after being appointed editor of the struggling *New York Daily News* in January 1997. Financially threatened and troubled by labor disputes, the *News* had adopted a superficial, celebrity-oriented approach at odds with the paper's history of hard-nosed news coverage. Under Hamill's guidance, however, there was a return to the paper's local emphasis, and a renewal of enthusiasm. According to Hamill, who hired a number of new young reporters (some bilingual), "everyone's going to want to read this paper and everyone in the newspaper business is going to want to work here." There was again an emphasis on good writing, investigative journalism, and coverage of minority communities. Hamill said "it's gotta be the immigrants and their children—that's our future." The paper was again making a profit and its prospects for future success were good.

Summary B

Essentially a celebration of veteran New York journalist Pete Hamill and *The New York Daily News*, this article summarizes the changes Hamill introduced after being appointed editor of the financially and otherwise troubled tabloid in January 1997. Little that's said here will be new to most readers of *The Columbia Journalism Review:* Hamill refocused the paper, abandoning its recently gossipy orientation and returning to its traditional emphasis on gritty local news and insider coverage of immigrant communities. Hamill's claim that "everyone's going to want to read this paper and everyone in the newspaper business is going to want to work here" typifies the article's fan-club tone. But there's little hard data presented to support such optimism. As the story itself admits, since the paper is privately owned, the publisher "doesn't have to furnish accompanying proof" to verify the claim that advertising linage is up by 50 percent and the *News* is again turning a profit. The article's credibility is further compromised by the fact that Hamill was fired shortly after its appearance.

Summary C

This three-page article discusses the career of veteran New York journalist Pete Hamill, focusing on his performance since being appointed editor of the financially troubled *New York Daily News* in January 1997. The article outlines several reforms Hamill introduced to return the paper to past prominence and suggests that his efforts were meeting with initial success.

■ EXERCISE 5.2

Write a 100-word descriptive summary of a major recent article from a periodical or Web site in your field of study or employment. Submit the article along with the summary.

■ EXERCISE 5.3

Write a 250-word informative summary of the same article mentioned in Exercise 5.2. Submit the article along with the summary.

■ EXERCISE 5.4

Write a 300-word evaluative summary of the same article mentioned in Exercises 5.2 and 5.3. Submit the article along with the summary.

■ EXERCISE 5.5

Write a 200-word informative summary of the plot of a recent episode of a favorite television show.

■ EXERCISE 5.6

Write a 250-word informative abstract of a term paper you have completed in the past for another course. Submit the term paper along with the abstract.

■ EXERCISE 5.7

Write an informative summary of an article from a popular periodical (for example, *Newsweek, Spin,* or *Sports Illustrated*). Make the summary no more than 20 percent as long as the article, and submit the article along with the summary.

EXERCISE 5.8

Summarize the instructor's lecture in one of your upcoming classes. Limit the summary to roughly 500 words.

EXERCISE 5.9

Write a seventy-five-word informative summary of an article from your local newspaper. Select an article at least 300 words long. Submit the article along with your summary.

EXERCISE 5.10

Write a fifty-word descriptive abstract of the sample report in Chapter 10.

6

Mechanism and Process/Procedure Descriptions

Learning Objective

Upon completing this chapter, you will be able to write clear, accurate mechanism and process/procedure descriptions.

The selective, well-organized presentation of significant details accompanied by visuals is an important form of workplace writing called description. An effective description enables a reader to accurately envision (and thereby better understand) an inanimate object, organism, substance, physical site, or activity. Obviously, the description's basic purpose is always to inform, but its specific uses are very broad. Like all workplace communications, the description is governed by its immediate purpose and the needs of the intended reader.

Depending on circumstances, a description may be quite short—like the one- or two-sentence caption accompanying a photograph of a house in a real estate agency's advertising pamphlet—or fully developed—like the highly detailed description a specialist would consult when preparing to service or repair the air-conditioning system in that same house. Accordingly, a description can stand alone or appear as part of a longer report or other document, such as a proposal, feasibility study, manual, or brochure.

This chapter focuses on the most common workplace applications: descriptions of mechanisms and of processes and procedures.

Mechanism Description

"Mechanism" refers to a tool, machine, or other mechanical device—usually with moving parts—designed to perform a specific kind of work. Naturally, your approach to writing a mechanism description (sometimes called a *device* description) will be influenced by your reader's needs. Your reader may require the description in order to identify, explain, advertise, display, package, ship, purchase, assemble, install, use, or repair the mechanism, and may possess technical understanding less highly developed than your own. You must therefore gear your description accordingly, taking all such considerations into account.

Your description will also be influenced by whether it is general or specific in nature. As the term suggests, a general description is accurate with respect to *all* such mechanisms, regardless of manufacturer, model, special features, or other variables. A general description of a camera, for example, would apply to all cameras—disposables, point-and-shoots, Polaroids, 35 mm, and others—by concentrating on the features common to all. A specific description, on the other hand, deals with one particular example of a mechanism—the Nikon One Touch Zoom camera, for instance—and emphasizes its particular and unique features.

Whether general or specific, nearly all mechanism descriptions include the following features:

- brief introduction defining the mechanism and explaining its purpose;
- precise description of the mechanism's appearance;
- list of the mechanism's major parts;
- explanation of how the mechanism works;
- one or more visuals that clearly depict the mechanism—photos and drawings, especially "exploded" and "cutaway" views, are best in this context (see Chapter 3);
- conclusion, sometimes incorporating information about the mechanism's history, availability, manufacturer, cost, etc.;
- list of outside sources of information, if any.

Obviously, the most *descriptive* parts of this sequence are the second, third, and fourth steps, which constitute the bulk of the text and necessitate certain procedures. You must decide, for example, what order of coverage to use in describing the object:

- top to bottom, or bottom to top;
- left to right, or right to left;
- inside to outside, or outside to inside;
- most important features to least important features, or least important features to most important features.

You should use specific, concrete wording—including the correct name of each part—and avoid vague, subjective expression that may result in misinterpretation. To a reader from a small town, for example, "a tall building" may mean any structure over two or three stories high. To a reader from New York or Chicago, however, "a tall building" means something quite different. Write exactly what you mean: "a fifty-story building" (or "a five-story building"). Similarly, don't write "a big long skinny thing"; instead, write "a seventeen-inch carbide spindle." Sometimes a little research is necessary to determine the correct specifications and terms, but this information is certainly available in dictionaries, encyclopedias, owner's and operator's manuals, merchandise catalogs, specialized reference works, on-line sources, and—for subject matter related to your major field—your textbooks. If you experience difficulty tracking down such information, any reference librarian can assist you.

Using the present tense and predominantly active verbs, explain completely how the mechanism looks. Mention all significant details: size, weight, shape, texture, and color. Identify what the mechanism is made of. Using familiar words and expressions like "above," "behind," "to the left of," "clockwise," "counter-clockwise," and the like, convey a clear sense of where the various parts are located in relation to each other, and how they interact. Evaluative comments can highlight the importance or significance of key details, as in these examples:

> The base of the machine is fitted with heavy-duty casters that, when unlocked, make it easy to move the machine from one location to another despite its great weight.
>
> The machine's simple design provides ready access to the motor, thus facilitating routine servicing and repairs.
>
> The housing's neutral color (standard on all models) allows it to blend in with the decor of most offices.

Use *analogy* and other forms of comparison to describe parts that are difficult to portray otherwise. We do this routinely when we employ letters of the alphabet in such expressions as "A-frame," "C-clamp," "T-square," and the like, or when we use terms like "wing nut," "needle-nose pliers," or "claw hammer." A little inventiveness will enable you to create fresh, original analogies to help the reader "see" what you're describing. But avoid vague, meaningless comparisons such as "the design resembles a European flag," which may be essentially accurate but, like the "tall building" example mentioned earlier, can be interpreted in many different ways.

Likewise, avoid analogies that depend on knowledge or understanding the reader may not possess. This is especially relevant now that the workplace is becoming increasingly diverse with respect to employees' national origins. Not everyone, for example, will understand sports analogies, especially those based upon lesser-known or exclusively American games. For example, the statement that a machine housing is "about as high and as wide as a lacrosse goal" would mean very little to a reader unfamiliar with that sport. Strive instead for analogies that most people can relate to. Restrict yourself to comparisons involving the universally known and recognized. To evoke an image of a wire surrounded by a layer of insulation, for example, you could liken it to the lead in a wooden pencil.

Ideally, of course, a visual will supplement the text to prevent misunderstanding. In mechanism description, the most useful visuals are photos and line drawings, especially cutaway and exploded views. As mentioned in Chapter 3, a familiar object such as a coin, a ruler, or even a human figure can be included within the picture to convey a sense of the mechanism's size, if that is not otherwise obvious. In a sense, this strategy is akin to analogy, which seeks to clarify by comparing the unknown to the known.

On the following pages are two mechanism descriptions for your consideration, commonplace examples that demonstrate basic formats, approaches, and strategies that can be adapted to actual workplace applications. Figure 6.1 is a general description of a conventional flush toilet, and Figure 6.2 is a specific description of an Ajax Super® ballpoint pen.

☑ Checklist Evaluating a Mechanism Description

A good mechanism description

___ opens with a brief introduction that defines the mechanism and explains its function;

___ fully describes the mechanism's component parts and how they interrelate;

___ is clear, accurate, and sufficiently detailed to satisfy the needs of the intended audience;

___ is well-organized, adopting the most logical sequence for the information;

___ employs helpful comparisons and analogies to clarify difficult concepts;

___ uses the present tense and an objective tone throughout;

___ uses clear, simple language;

___ concludes with a brief summary;

___ employs effective visuals (photographs and line drawings—exploded and/or cutaway views) to clarify the text;

___ contains no typos or mechanical errors in spelling, capitalization, punctuation, and grammar.

A Conventional Flush Toilet

Introduction

A flush toilet is a mechanical device for the sanitary disposal of bathroom waste matter.

Appearance

Typically made of white porcelain, the toilet consists of two basic components: an oval, water-filled bowl with a hinged plastic seat and cover; and a rectangular water tank positioned directly behind the bowl and fitted with a metal or porcelain flush/trip handle on the upper left-hand side of its front surface.

The rim of the bowl is approximately 15" from the floor. The bowl is approximately 15" across at its widest point. The tank is approximately 20" wide, 14" high, and 8" deep. The tanks of older models held approximately 3.5 gallons of water, while current models hold 1.5 gallons.

The seat and cover on newer models are easily removable to facilitate cleaning. The tank lid is also removable, to allow adjustment and repair of the mechanical parts housed inside.

Major Parts

Virtually all the major parts are inside the tank:

- lift wires
- guide arm
- tank ball
- valve seat
- float ball
- float arm
- ball-cock assembly
- inlet tube
- filler tube
- overflow tube

FIGURE 6.1 Example of a General Mechanism Description, page 1

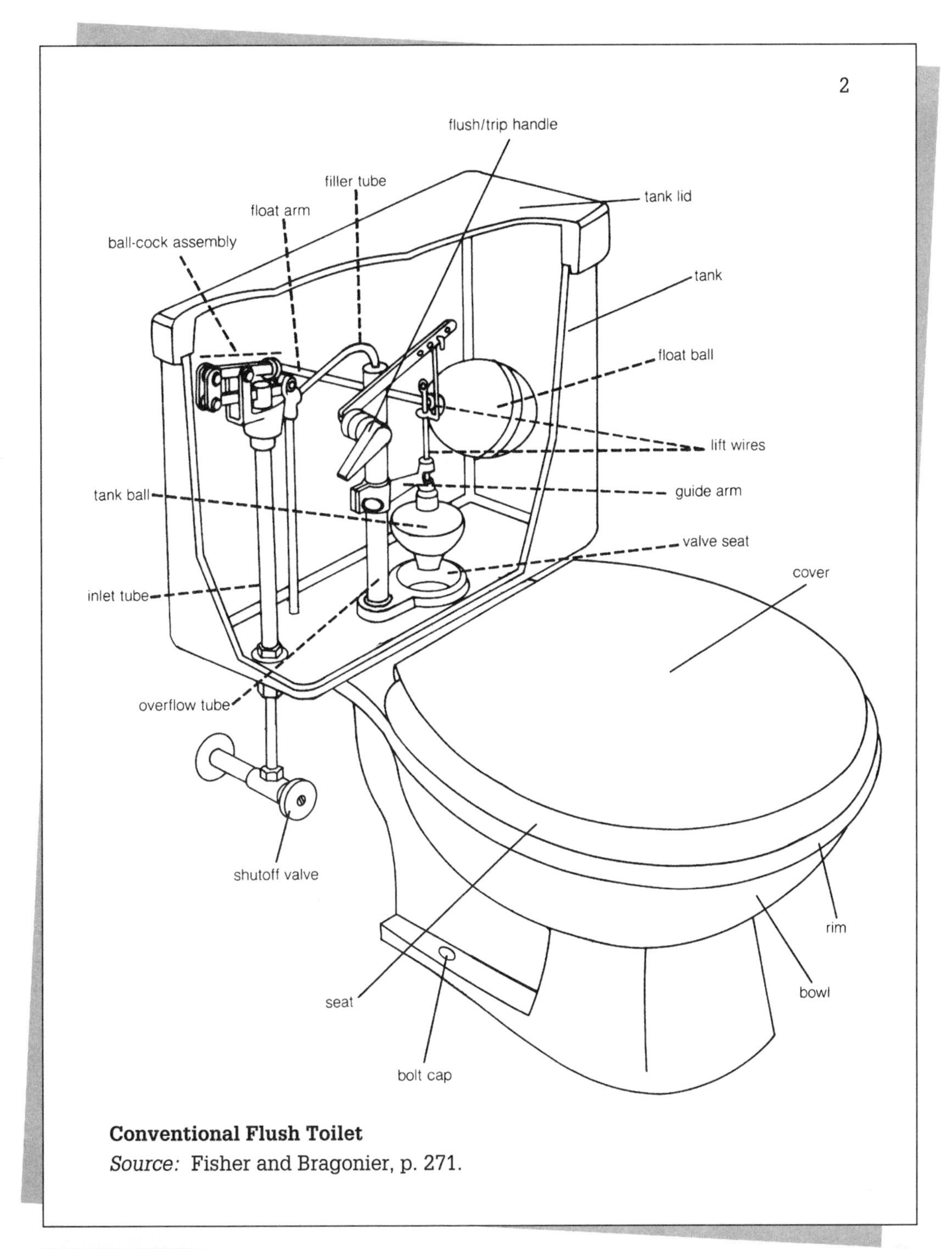

Conventional Flush Toilet

Source: Fisher and Bragonier, p. 271.

FIGURE 6.1 Example of a General Mechanism Description, page 2

3

Operation

1. Depressing the flush/trip handle lifts the tank ball from the valve seat, allowing water from the tank to empty into the bowl and carry away waste through a pipe in the floor. When the tank is nearly empty, the ball falls back into the seat and stops the flow.

2. As the water level in the tank falls, so does the float ball. This causes the float arm to open a valve in the ball-cock assembly, letting new water into the inlet tube and allowing both the tank and the bowl to refill through the filler tube.

3. As the water level in the tank rises, so does the float ball. Eventually the float arm closes the valve in the ball-cock assembly, shutting off the water until the next flush. In the event of malfunction, excess water escapes through the overflow tube.

Conclusion

For more than 400 years, people have disposed of waste matter efficiently and hygienically by means of the flush toilet. Sir John Harrington, a godson of Queen Elizabeth I of England, invented the flush toilet in 1589. A valve released a flow of water from a cistern tank into a flush pipe. But the valve tended to leak, creating a nearly continuous trickle of water into the bowl.

Chelsea plumber Thomas Crapper's "Valveless Water-Waste Preventer," introduced at the Health Exhibition of 1884, solved this problem by incorporating the float-ball principle in use ever since.

Today the leading manufacturers of flush toilets are Kohler and American Standard. A basic model retails for approximately $150, but special features such as brass hardware, elongated bowls, lined tanks, and designer colors can boost the price to $1,000 or more.

Sources

Fisher, David, and Reginald Bragonier, Jr. *What's What: A Visual Glossary of the Physical World.* Maplewood, NJ: Hammond, 1981.

Macaulay, David. *The Way Things Work.* Boston: Houghton Mifflin, 1988.

Reyburn, Wallace. *Flushed With Pride.* London: MacDonald, 1969.

FIGURE 6.1 Example of a General Mechanism Description, page 3

An Ajax Super® Ballpoint Pen

Introduction

A ballpoint pen is a common writing implement used in homes, schools, and offices—indeed, in virtually every setting where writing is done—all over the developed world.

Appearance

Although an inch or two shorter, the ballpoint pen looks much like a pencil. Some disposable ballpoint pens are of unified construction, but most consist of a tapered barrel and a somewhat shorter cap, which—fitted together—give the pen its elongated shape and allow it to be disassembled for the purpose of replacing the tubular ink reservoir inside. Most ballpoint pens have a tension clip mounted on the cap, to allow the pen to be secured in the owner's shirt pocket when not in use. (The clip also prevents the pen from rolling off inclined surfaces.) In addition, there is a small push-button that protrudes from the top of the cap, where the eraser would be on a pencil. This activates the inner workings, allowing the tiny, socket-mounted, rolling ballpoint writing tip at the end of the reservoir to be extended or retracted.

The visible parts of the pen can be made entirely of metal, although many pens are both metal and plastic (metal cap, clip, and button; plastic barrel). Typically, however, it is made mostly of plastic. Metal surfaces can be shiny or textured. Plastic parts are manufactured in a wide range of colors, and some ballpoint pens are two-tone (one color for the cap, another for the barrel).

The Ajax Super model described here is exactly 5" long, with a 3/8" diameter at its widest point, where the cap and barrel join. Both cap and barrel are made of bright red styrene plastic. The pocket-clip and the push button are made of chromed metal.

Major Parts

Exterior Parts	Interior Parts
• push-button	• plunger mechanism
• cap	• ink reservoir
• pocket clip	• spring
• barrel	• ballpoint

FIGURE 6.2 **Example of a Specific Mechanism Description, page 1**

2

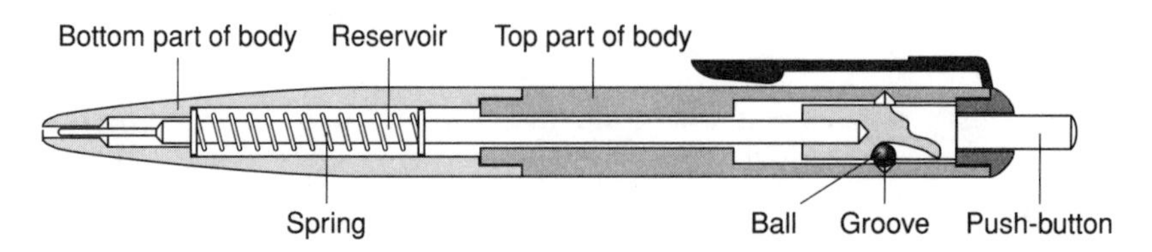

Ajax Super® Ballpoint Pen
Source: How Things Work, p. 313.

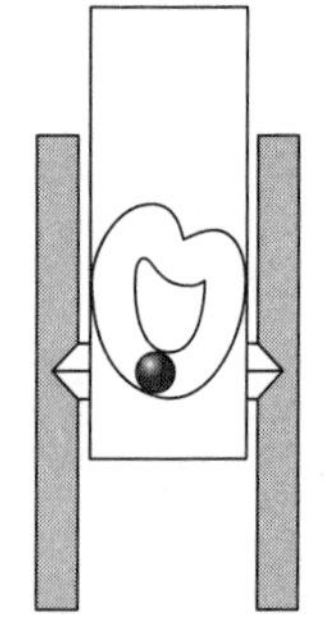

Initial position: ball in bottom holding point

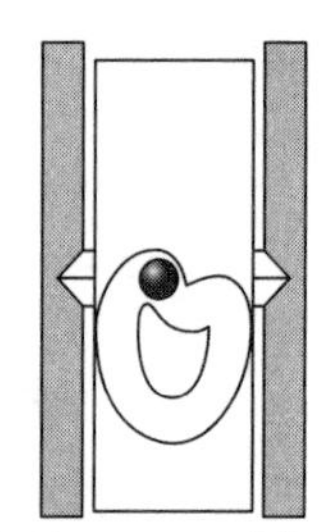

Button pressed: ball in recess

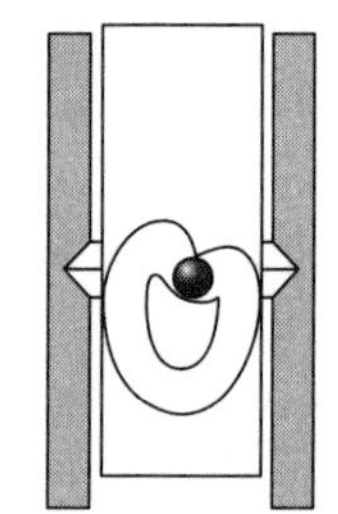

Position of ball determines position of reservoir

Operating Principle
Source: How Things Work, p. 313.

Operation

1. Depressing the push-button causes the interior plunger mechanism to activate, thereby forcing the ink cartridge downward, locking it in place, and causing the ballpoint writing tip to extend through the opening at the tapered end of the barrel. Plunger mechanisms vary from pen to pen, with some more complicated than others.

FIGURE 6.2 **Example of a Specific Mechanism Description, page 2**

3

2. The Ajax Super boasts a rather simple mechanism employing a ball-catch. When the push-button is depressed, a tiny ball rotates in a heart-shaped cam recess in the plunger's cylindrical shaft. After the push-button is depressed, the pressure of the spring forces this ball into the topmost holding-point within the recess and prevents its return, thus locking the reservoir and the ballpoint writing tip in the extended position.
3. As the ballpoint rolls across the writing surface, ink is drawn from the reservoir and onto the ball, which then deposits it.
4. When the push-button is again depressed, the mechanism is disengaged and the spring causes the cam ball to return to its original holding point within the recess, allowing the reservoir and its ball-point tip to retract into the barrel. The now-stationary ballpoint seals the reservoir, preventing unwanted discharge of ink while the pen is not in use.

Conclusion

The ballpoint pen has evolved greatly since its introduction in the 1940s. Early models were notoriously messy and unreliable, with leakage a constant problem. Indeed, the ballpoint pen was initially banned in many public schools. Since then, however, manufacturers such as Parker, Papermate, Cross, Bic, and others have perfected the device, which is now the most popular handheld writing instrument, having almost completely supplanted the traditional fountain pen. Many varieties of ballpoint pens are available, with prices ranging from less than a dollar for unmechanized, disposable plastic models to literally hundreds or even thousands of dollars for high-prestige, brand-name instruments made of gold or silver.

Sources

Fisher, David, and Reginald Bragonier, Jr. *What's What: A Visual Glossary of the Physical World.* Maplewood, NJ: Hammond, 1981.

How Things Work: The Universal Encyclopedia of Machines. London: Paladin, 1972.

Macaulay, David. *The Way Things Work.* Boston: Houghton Mifflin, 1988.

FIGURE 6.2 **Example of a Specific Mechanism Description, page 3**

Exercises

■ EXERCISE 6.1

As discussed in the text and mentioned on the checklist, a mechanism description should open with a brief introduction that defines the mechanism and explains its function. To acquire some practice with this task, write introductions for three of the following metering devices:

- ammeter
- barometer
- odometer
- psychrometer
- spectrophotometer

■ EXERCISE 6.2

Write a general mechanism description of a common device used in your field of study or employment—for example, welding torch, compass, stethoscope.

■ EXERCISE 6.3

Write a specific mechanism description of a common kitchen appliance—for example, toaster, blender, coffee-maker.

■ EXERCISE 6.4

Write a specific mechanism description of a piece of sporting equipment—for example, ski boot, baseball or softball glove, golf club.

■ EXERCISE 6.5

Write a general mechanism description of a tool or other piece of equipment commonly used in automobile repair—for example, creeper, ratchet wrench, bumper jack.

■ EXERCISE 6.6

Write a specific mechanism description of a Swiss Army knife.

EXERCISE 6.7

Write a general mechanism description of a common piece of office equipment—for example, stapler, locking file cabinet, telephone answering machine.

EXERCISE 6.8

Consult some textbooks, periodicals, and Web sites devoted to your field of study or employment and find ten examples of the creative use of analogy. Write a memo report to your instructor, discussing your findings.

EXERCISE 6.9

Consult some textbooks, periodicals, and Web sites devoted to your field of study or employment and find five examples of effective visuals that successfully clarify the appearance or workings of mechanisms. Write a booklet report to your instructor, in which you discuss your findings. Be sure to include copies of the visuals.

EXERCISE 6.10

As an exercise in precise writing, create a detailed description of the following illustration. When you think your description is as clear as possible, give it to a friend, relative, or co-worker who has not seen the illustration and ask the person to read the description and reproduce the illustration. If the resulting drawing is not an exact replication of the original, it's probably because your description was not 100 percent clear. Discuss the results with your "artist," to determine exactly what was misleading.

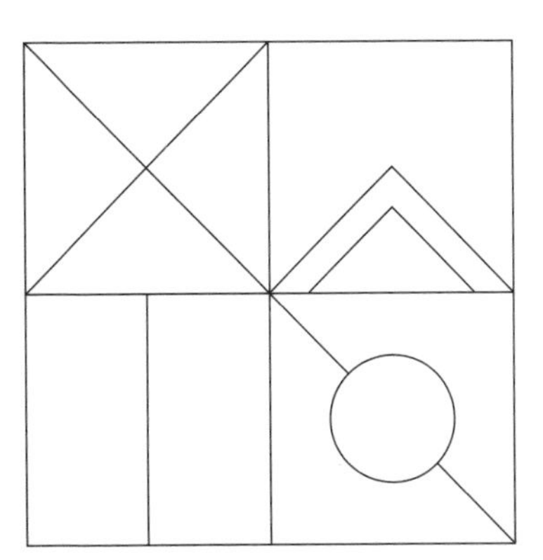

Process/Procedure Description

Process description is similar to mechanism description. As the term suggests, however, a process description focuses not on an object but on an unvarying series of events producing a predictable outcome. Sometimes referred to as process *analysis,* it enables the reader to understand—but not necessarily to create—a particular process. Indeed, processes as strictly defined are natural phenomena governed by physical laws, and are therefore beyond the scope of deliberate human involvement. Photosynthesis, for example, is a process, as is continental drift or gene mutation. A more flexible definition of process would include human-controlled activities, such as data processing.

A predetermined series of events that occurs under human control is most typically referred to as a procedure. A manufacturing operation such as injection molding, for example, is a procedure, as is balancing a checkbook or measuring a pulse rate. Hence, if process description most commonly answers the question, How and why does X *happen*? procedure description always answers the question, How and why is X *done*? These are useful distinctions.

A procedure description differs sharply from instructions, which are covered in Chapter 7, because it enables the reader to understand the procedure but not necessarily to perform it. The purpose of procedure description, like that of process description, is simply to inform. Indeed, both kinds of description are treated together here because they are so similar. In addition to their common purpose, they are also structured the same. Furthermore, both resemble mechanism description. Like mechanism description, both can be presented in a variety of contexts and usually include certain features similar to those of mechanism description:

- brief introduction explaining the nature, purpose, and importance of the process or procedure;
- explanation of the natural forces and (if any) the materials, tools, and other equipment involved;
- stage-by-stage explanation of how the process or procedure occurs, with all necessary details;
- one or more visuals for purposes of clarification—flow charts are often useful in this context (see Chapter 3);
- brief conclusion;
- list of outside sources of information, if any.

As in other kinds of workplace writing, use *parallel structure* when phrasing the stage-by-stage explanation. Do not say, for example,

1. Customer drops off the clothing.
2. Clothing is sorted and marked.
3. Clothes you want washed go to washing machines, and clothes to be dry-cleaned go to dry-cleaning machines.

Instead, be consistent. Phrase the description as follows:

1. Clothing is dropped off.
2. Clothing is sorted and marked.
3. Clothing to be washed is sent to washing machines, and clothing to be dry-cleaned is sent to dry-cleaning machines.

The revision is better because the emphasis is now on the same thing ("clothing") in each of the three stages, and the verbs are now all passive. Since the focus in this kind of description is on the process or procedure itself rather than on a human agent, this is one of the few situations in which passive verbs (see Appendix A) rather than commands or other active constructions are preferable. Indeed, when there is no human agent—as in a sentence like "steam is created as water evaporates"—passive voice is the only way to express the idea smoothly.

As with mechanism description, the level of detail and technicality in process/procedure descriptions will depend on the needs and background of the audience. If a process description is especially detailed, it may even include sections of mechanism description within it. In any case, the sequential, cause-and-effect, action-and-reaction nature of the information must be conveyed clearly to the reader, to ensure understanding. Do not simply say, for example:

1. X happens.
2. Y happens.
3. Z happens.

Instead, use transitions to reveal the relationships among the stages of the process or procedure, like this:

1. X happens.
2. Meanwhile, Y happens.
3. As a result, Z happens.

Just as you might when writing a mechanism description, you can use analogies and familiar comparisons to convey difficult concepts involved in the process or procedure you are describing. Visuals can also be quite helpful. In general, the most appropriate kind of visual to accompany a process/procedure description is the flow chart, which by nature is intended to illustrate activity through a series of stages (see Figure 3.6). Sometimes, though, a more pictorial touch is helpful, as in the process description in Figure 6.3. Conversely, a simple, step-by-step narrative can be sufficiently clear by itself, as demonstrated by the procedure description in Figure 6.4.

Evaluating Process/Procedure Description

A good process/procedure description

___ opens with a brief introduction that identifies the process or procedure and explains its function;

___ fully describes the process or procedure, providing all details as to when, where, why, and how it occurs;

___ is clear, accurate, and sufficiently detailed to satisfy the needs of the intended audience;

___ is well organized, adopting the most logical sequence for the information;

___ employs helpful comparisons and analogies to clarify difficult concepts;

___ uses the present tense and an objective tone throughout;

___ uses clear, simple language;

___ concludes with a brief summary;

___ employs effective visuals (flow charts and organizational charts, for example) to clarify the text;

___ contains no typos or mechanical errors in spelling, capitalization, punctuation, and grammar.

Erosion and Sedimentation

Introduction

Broadly defined, erosion and sedimentation constitute the two-step process whereby natural forces carry away soil, rock, and other surface materials and transport them to new locations on the earth's surface.

This process causes a gradual and ongoing reconfiguration of the earth's topography. Over the centuries it has created riverbeds, waterfalls, rapids, etc.

Natural Forces

The main natural forces involved in erosion and sedimentation are water, wind, and ice.

Process

WATER—According to Professor W.H. Matthews III, of Lamar University, "streams cause more erosion than all other geological agents combined." The forceful scouring action of rushing water in rivers and estuaries, abetted by abrasive sediment churned up from beneath the surface and carried along, erodes the banks, creates more sediment, and eventually widens the channel. This process is accelerated during flooding, which increases the volume and therefore the velocity of the streamflow. In coastal regions, seawater creates much the same effect. Ocean waves—aided by gravity, wind, and rain—slowly break down bedrock and reshape shoreline cliffs. Just as freshwater carries abrasive sediment that enhances the water's erosive capability, seawater contains grit created by subaqueous erosion resulting from turbulent currents.

FIGURE 6.3 **Example of a Process Description, page 1**

2

U.S. NATIONAL PARK SERVICE

Landscape Arch in Utah, 291 feet (88 meters) long, was chiseled from bedrock by wind-driven sand.

Source: Encyclopedia Americana, p. 556.

ICE—As water freezes, it expands with great force, breaking apart rocks whose crevices it occupies. Glaciers too have played a major role in erosion. As a glacier moves slowly along, it abrades the earth's surface, partly because of debris quarried from the bedrock by the glacier and embedded in the ice along its base. North America's Great Lakes, in fact, were formed by such glacial activity.

WIND—Wind is a factor primarily in arid, sandy regions where the ground is unprotected by vegetation or other cover. In such areas whole landscapes of hills and valleys can result, as on the planet Mars and in several locations on earth.

FIGURE 6.3 Example of a Process Description, page 2

3

Conclusion

The process of erosion and sedimentation is positive because in addition to creating some spectacular geophysical phenomena (see illustration) it contributes to the formation of new soil and causes rich deposits on valley floors and at the mouths of rivers. Its harmful effects include the depletion of existing topsoil and fertilizer, and the clogging of drainpipes and reservoirs. Although significant, the process is but one of several related influences upon the earth's topography. Others include tectonic activity and surface earth movement (landslides, for example). In recent years more attention is being paid to the role of human ecology, particularly the consequences of such practices as quarrying, strip mining, and large-scale deforestation.

Sources

"Erosion." *New Encyclopædia Britannica: Micropædia,* 1992 ed.

Garner, H.F. "Erosion and Sedimentation." *Academic American Encyclopedia,* 1993 ed.

Laflen, John M. "Erosion." *World Book Encyclopedia,* 1986 ed.

Matthews, William H. "Erosion." *Encyclopedia Americana,* 1994 ed.

FIGURE 6.3 **Example of a Process Description, page 3**

Problem Resolution at the Conover Corporation

Introduction

The Conover Corporation is committed to maintaining a climate of open, honest communication between employees and their immediate supervisors, for the purpose of resolving problems that may undermine morale and thereby hinder the achievement of corporate goals. To ensure that problems are resolved promptly and fairly, a procedure has been established, subject to periodic review.

Procedure

1. In the event of a job-related problem, the employee requests a brief conference with the immediate supervisor, to agree upon a mutually acceptable time, date, and place to meet for discussion.
2. The employee and the supervisor meet. If possible, the supervisor resolves the problem. If additional information, assistance, or time is required, the supervisor arranges a second meeting with the employee, again by mutual agreement.
3. If dissatisfied with the outcome, the employee asks that the matter be referred to the next higher level of supervision. The supervisor tries to arrange this meeting within two working days of the employee's request, and summarizes in a memo report the nature of the problem and the initial response given to the employee. Prior to this meeting, copies of the report are given to the employee, the next higher level of supervision, and the Human Resources Department. At the meeting, the employee may explain the problem informally (in conversation, which will result in an oral response) or formally (in the form of a memo report, which will result in a written response).
4. If still dissatisfied, the employee may ask that the matter be referred to the next higher level of management; if necessary, the procedure is repeated until it reaches the point at which a representative of the Human Resources Department joins in the review, and the matter is finally resolved.

Conclusion

Employees are encouraged to use this procedure without fear of reprisal or penalty. Records of problem review procedures are not entered into employees' personnel files, nor are they a factor in performance reviews. Such records are maintained in a separate file in the Human Resource Office, but only for reference, to ensure consistency in the handling of other such situations.

FIGURE 6.4 Example of a Procedure Description

Exercises

EXERCISE 6.11

Identify each of the following as either a process or a procedure.

- alphabetizing and filing
- corrosion
- intubation
- solving a quadratic equation
- vaporization

EXERCISE 6.12

As discussed in the text and mentioned on the checklist, a process/procedure description should open with a brief introduction that defines the process or procedure and explains its function. Write such introductions for three of the items in Exercise 6.11.

EXERCISE 6.13

Write a description of a process related to your field of study or employment—for example, pulmonary arrest, cell regeneration, or food spoilage.

EXERCISE 6.14

Write a description of a procedure related to your field of study or employment—for example, interviewing, stock rotation, or safety check.

EXERCISE 6.15

Write a description of a procedure that occurs in the world of sports—for example, the annual National Basketball Association draft, the way in which points are awarded in diving competitions, or the way the starting line is organized at the Boston or New York Marathon.

■ EXERCISE 6.16

Write a process description of how cigarette smoking affects the lungs.

■ EXERCISE 6.17

Write a process description of how body rot forms on a motor vehicle.

■ EXERCISE 6.18

Write a procedure description of how a dealer determines the market value of a collectible—for example, baseball cards, coins, comic books, figurines, or postage stamps.

■ EXERCISE 6.19

Write a process description of how spontaneous combustion occurs.

■ EXERCISE 6.20

Consult some textbooks, periodicals, and Web sites devoted to your field of study or employment and find five examples of effective visuals that successfully clarify processes or procedures. Write a booklet report to your instructor, in which you discuss your findings. Be sure to include copies of the visuals.

7

Instructions

Learning Objective

Upon completing this chapter, you will be able to write clear instructions that will enable the reader to perform a given procedure without unnecessary difficulty or risk.

Instructions serve a wide variety of functions. You might write instructions for co-workers to enable them to install, operate, maintain, or repair a piece of equipment, or to follow established policies such as those explained in employee handbooks. You might also write instructions for customers or clients to enable them to assemble, use, or maintain a product (as in owner's manuals, for example) or to follow mandated guidelines. As in procedure description writing, the broad purpose of instructions is to inform. The more specific purpose, however, is to enable the reader to *perform* a particular procedure rather than simply understand it.

Clearly, instructions must be closely geared to the needs of the intended reader. The level of specificity will vary greatly, depending on the procedure's complexity and context, and the reader's level of expertise or preparation. Computer documentation intended for a professional programmer, for example, is very different from documentation written for someone with little experience in such matters. Obviously, audience analysis is crucial to writing effective instructions.

Instructions

Just as there are general and specific mechanism descriptions, there are general and specific instructions. The differences are essentially the same for both types of workplace writing. General instructions explain how to perform a generic procedure—trimming a hedge, for example—and can be adapted to individual situations. Specific instructions explain how to perform a procedure under conditions involving particular equipment, surroundings, or other such variables—operating an 18-inch Black & Decker Auto Stop electric hedge trimmer, for instance.

Like other kinds of workplace writing, instructions appear in diverse contexts, from brief notes such as the reminder to "Close cover before striking" that appears on every matchbook, to lengthy manuals and handbooks. Most instruction writing, however, regardless of context, follows a basic format that resembles that of a recipe in a cookbook. This format includes the following basic features:

- brief introduction explaining the purpose and importance of the procedure;

Note: an estimate of how much time will be required for completion of the procedure may be included as well as any unusual circumstances that the reader must keep in mind *throughout,* such as safety considerations.

- list of materials, equipment, tools, and skills required, enabling the reader to perform the procedure uninterrupted;
- the actual instructions: a numbered, step-by-step, detailed explanation of how to perform the procedure;
- in most cases, one or more visuals for clarification;
- brief conclusion;
- list of outside sources of information, if any.

Although they *look* very easy, instructions are actually among the most difficult kinds of writing to compose. The slightest error or lapse in clarity can badly mislead—or even endanger. Instructions should always be read in their entirety before the reader attempts the task. Many readers, however, read the instructions a bit at a time, "on the fly," while already performing the procedure. This puts an even greater burden on the writer to achieve standards of absolute precision and clarity.

The best approach is to use short, simple commands that start with a verb, arranged in a chronologically numbered list. This will enable the reader to follow the directions without confusion, and will also foster consistent, action-focused wording, as in this example:

1. Push the red "On" button.
2. Insert the green plug into the left-hand outlet.
3. Push the blue "Direction" lever to the right.

Notice that instructions are not expressed in "recipe shorthand." Small words such as "a," "an," and "the," which would be omitted in a recipe, are included in instructions.

Although it is usually best to limit each command to one action, sometimes closely related steps can be combined to prevent the list from becoming unwieldy. Consider the following example.

1. Hold the bottle in your left hand.
2. Twist off the cap with your right hand.

These steps should probably be combined as follows:

1. Holding the bottle in your left hand, twist off the cap with your right hand.

When the procedure is very complicated and therefore requires a long list, a good strategy is to use subdivisions under major headings, like this:

1. Prepare the solution:
   a) pour two drops of the red liquid into the vial,
   b) pour one drop of the blue liquid into the vial,
   c) pour three drops of the green liquid into the vial,
   d) cap the vial,
   e) shake the vial vigorously for ten seconds.
2. Pour the solution into a beaker.
3. Heat the beaker until bubbles form on the surface of the solution.

If two actions *must* be performed simultaneously, however, present them together. For example, do not write something like this:

1. Push the blue lever forward;
2. Before releasing the blue lever, push the red button twice;
3. Release the blue lever.

Instead, write this:

1. While holding the blue lever in the forward position, push the red button twice;
2. Release the blue lever.

Remember that what may seem obvious to you is not necessarily apparent to the reader. Include all information and provide the reason for each step if the reason is important, as in this example:

> Keep the clutch pedal depressed whenever the car is stopped while in gear, <u>or the car will buck forward and stall</u>.

Often, as in the above instance, the reason for doing something a certain way is to prevent inconvenience or malfunction. But if a serious danger exists, you must alert the reader by using LARGE PRINT, underlining, **boldface**, or some other attention-getter, *before* the danger point is reached. There are three kinds of alerts, as follows.

- ***Warning:*** Alerts the reader to the possibility of injury or death, as in this example:

  WARNING: To avoid the risk of injury or death by electrocution, you must turn the power off before removing the cover plate.

- ***Caution:*** Alerts the reader to the possibility of equipment damage, as in this example:

  CAUTION: To prevent damage to the blades, do not permit any metal object to enter the cutter.

- ***Note:*** Alerts the reader to information that will make the procedure easier or more efficient, as in this example:

  NOTE: Lubricate the axle now, because it will be much more difficult to reach after the housing is in place.

Including such advisories in instructions is now more important than ever, not only to prevent damage and injury but also to minimize legal liability in the event of a mishap. The frequency of product liability lawsuits has skyrocketed in recent years, with the great majority of such suits alleging not that the products themselves are defective, but that manufacturers have failed to provide sufficient warnings about dangers inherent to their use. As a result, even obvious precautions must be spelled out in very explicit terms. The McDonald's Corporation, for example, now includes the phrase "Caution: Contents Hot" on its coffee cups, in response to losing $2.7 million in a lawsuit to a customer who was scalded by a spilled beverage and sued for damages. Another recent example is the superhero costume that actually carried the following message on its packaging:

> PARENT: Please exercise caution—FOR PLAY ONLY. Mask and chest plates are not protective: cape does not enable user to fly.

These are, of course, extreme cases. But they underscore the importance of providing ample warning in your instructions about any potential hazard. Similarly, if malfunction can occur at any point, you should explain corrective measures, as in these examples:

> If the belt slips off the drive-wheel, disengage the clutch.
>
> If any of the solution splashes into the eyes or onto exposed skin, wash immediately with cold water.
>
> If the motor begins to whine, immediately turn off the power.

Frequently the conclusion to a set of instructions will take the form of a "troubleshooting" section in which possible causes of difficulty are identified along with remedies. This example is from Marine Midland Bank's instructions on how to balance a checkbook:

> Subtract Line 4 from Line 3. This should be your present register balance. If not, the most common mistakes are either an error in arithmetic or a service charge not listed in your register. If you need further assistance, please bring this statement to your banking office.

Sometimes the troubleshooting guide will be in the form of a three-column "fault table" such as the following example, which appeared in an automobile owner's manual.

Symptom	**Probable Cause**	**Solution**
Starter motor won't work	1. Loose connections 2. Weak battery 3. Worn-out motor	1. Tighten connections 2. Charge battery 3. Replace motor
Starter motor works but engine won't start	1. Wrong starting procedure 2. Flooded engine 3. No fuel 4. Blown fuse 5. Ignition defect, fuel-line blockage	1. Correct procedure 2. Wait awhile 3. Refuel 4. Replace fuse 5. Contact dealer
Rough idle, stalling	1. Ignition defect, fuel-line blockage	1. Contact dealer

Another helpful feature of good instructions is the use of effective visuals. Photographs, line drawings (especially cutaway and exploded views), and flow charts can clarify concepts that might otherwise be difficult to understand. As explained in Chapter 3, however, you must choose the right type of visual for each situation, and this is certainly true with respect to writing instructions. Different kinds of instructions are best illustrated by different kinds of visuals. A precise operation involving the manipulation of small parts, for example, may best be rendered by a close-up photograph or line drawing of someone's hands performing the operation. Instructions emphasizing the correct sequence for the steps in a less delicate task, on the other hand, may best be illustrated by a conventional flow chart.

Increasingly, instructions designed to accompany products feature visuals alone or with minimal text. The principal reason for this development is that manufacturers wish to target the broadest possible market by accommodating consumers from various countries and cultures. This trend is likely to grow as we move ever closer to a global economy.

A good way to determine the effectiveness of a set of instructions you have created—whether with text and visuals or with visuals alone—is to field test them by observing while someone unfamiliar with the procedure attempts to perform it using your directions. For the test to be valid, however, you must resist the temptation to provide verbal assistance if the person expresses uncertainty. This will enable you to detect any unclear sections within the instructions, and to determine the cause of the confusion. Another effective test is to ask someone who *is* familiar with the procedure to critique your instructions. Even better, subject your instructions to both forms of evaluation.

The following examples illustrate the principal types of instructions, general and specific. Figure 7.1 explains how to perform a very common procedure: the changing of a flat tire. Figure 7.2 shows how to assemble a specific product: a Weber charcoal barbecue kettle. Note that although the Weber instructions rely almost exclusively upon visuals, the reader is referred to the safety measures explained in the accompanying Owner's Guide.

How to Change a Flat Tire

Introduction

Nearly every motorist experiences a flat tire sooner or later. Therefore, you should know what to do in such a situation. Changing a flat is fairly simple, but the correct procedure must be followed to prevent injury or vehicle damage.

Tools and Equipment

To change a tire you will need the following:

- flares (6–10)
- flashlight (if at night)
- spare tire, mounted
- tire pressure gauge
- wheel blocks (2)
- jack handle
- screwdriver
- penetrating oil
- lug wrench
- wide board
- automobile jack
- rubber mallet

Procedure

As soon as you realize you are developing a flat, leave the roadway and drive your vehicle onto the road shoulder. Park as far from the road and on as flat and level a surface as possible. Turn off the engine and activate the hazard warning flashers. Put the transmission in "Park." (Put a manual transmission in reverse.) Set the parking brake. Get all passengers out of the vehicle and have them stand well away from traffic and clear of the vehicle. Raise the hood to warn other motorists and to signal that you may need help. Now you are ready to assess the situation.

WARNING: DO NOT ATTEMPT TO CHANGE A TIRE IF YOUR VEHICLE IS ON AN INCLINE OR SLOPE, OR IF ONCOMING TRAFFIC IS DANGEROUSLY CLOSE TO YOUR VEHICLE. UNDER THESE CONDITIONS, WAIT FOR PROFESSIONAL ASSISTANCE.

1. If conditions are acceptable, open the trunk and set up flares behind and in front of the vehicle, to alert other motorists.

FIGURE 7.1 Example of General Instructions, page 1

2. Remove the spare and other equipment from the trunk.
3. Using the jack handle, pry off the hubcap. If the jack handle is not satisfactory for this, use the screwdriver.
4. Using the lug wrench, loosen (but do not remove) the lug nuts. Nearly all lug nuts are loosened by turning counter-clockwise. (If the nuts have left-hand threads and are therefore loosened clockwise, there will be an "L" on the lug bolt.) If the nuts are too tight, apply the penetrating oil, wait a few minutes, and try again.
5. Assemble and position the jack. Since there are several kinds, you must consult your owner's manual for proper assembly and use. If the ground is soft, put the wide board under the jack base to stabilize it.
6. Put the wheel blocks in front of and behind the tire diagonally opposite the one being changed, to minimize the risk of the vehicle rolling off the jack. (See Figure 1)
7. Raise the vehicle until the flat tire is just clear of the ground. ALWAYS REMOVE THE JACK HANDLE WHEN NOT IN USE.

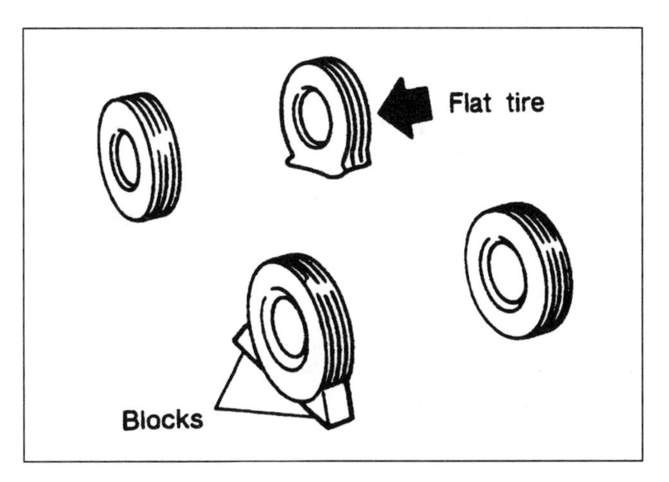

Figure 1 Position of Wheel Blocks

FIGURE 7.1 Example of General Instructions, page 2

8. Remove the lug nuts by hand. NEVER USE THE LUG WRENCH WHILE THE VEHICLE IS JACKED UP. If the lug nuts will not come off by hand, lower the car, further loosen them with the wrench, jack the car back up, and then remove them by hand. (Like removing the jack handle, this will minimize the risk of accidentally dislodging the jack—a dangerous error!) Put the lug nuts into the hubcap for safekeeping.
9. Remove the flat.
10. Roll the spare into position and put it on the wheel by aligning the holes in the spare's rim with the lug bolts on the wheel. You may have to jack the vehicle up a bit more to accomplish this, as the properly inflated spare will have a larger diameter than the flat.
11. Holding the spare firmly against the wheel with one hand, use your other hand to replace the lug nuts as tightly as possible. Again, DO NOT USE THE WRENCH.
12. Lower the vehicle. Now you may use the wrench to fully tighten the nuts. To ensure that the stress is distributed evenly, tighten the nuts in the proper sequence shown in Figure 2.

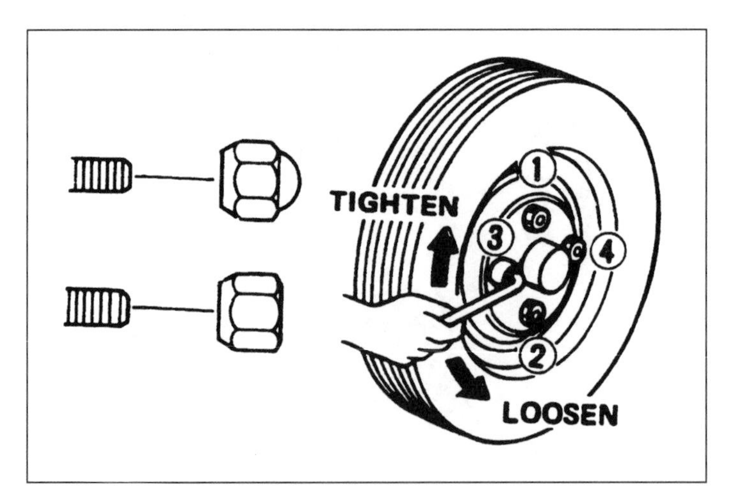

Figure 2 Lug Nut Sequence

FIGURE 7.1 Example of General Instructions, page 3

4

13. Replace the hubcap, checking to ensure that the tire valve is correctly positioned, protruding through the hole in the hubcap. You may need to tap the hubcap into place with the rubber mallet.

14. Put the flat, jack, and other tools into the trunk.

Conclusion

If correct procedure is followed, most motorists are able to change a flat tire successfully. The only difficult part of the task is the removal of the lug nuts, which does require some physical strength. As explained earlier, however, penetrating oil will help, as will a long-handled lug wrench, which provides greater leverage. A common practice is to use one's foot to push down on the wrench handle.

Sources

Nissan Pulsar NX Owner's Manual. Tokyo, Japan: Nissan Motor Co., Ltd., 1988.

Pettis, A.M. *Monarch Illustrated Guide to Car Care.* New York: Simon & Schuster, 1977.

Reader's Digest Complete Car Care Manual. Pleasantville, NY: Reader's Digest Association, 1981.

FIGURE 7.1 **Example of General Instructions, page 4**

weber BAR-B-KETTLE™ GRILL

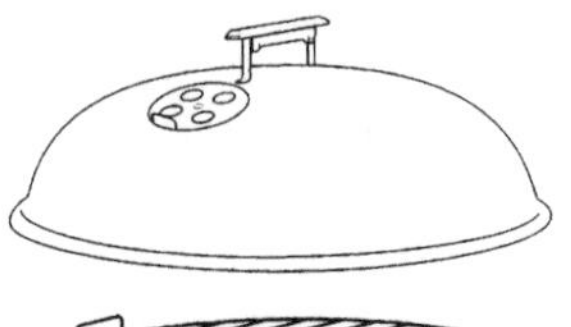

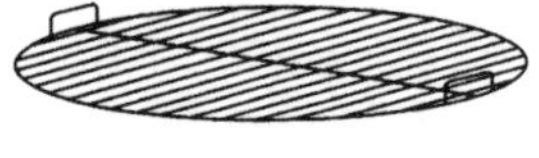

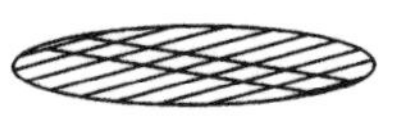

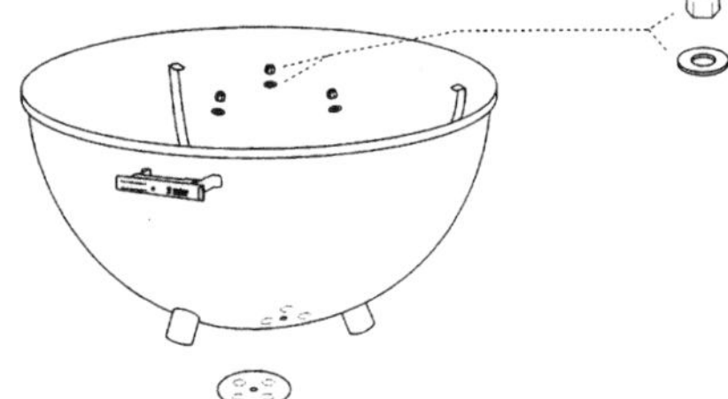

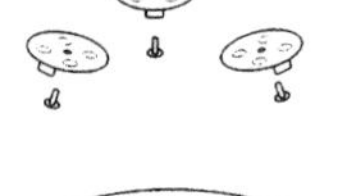

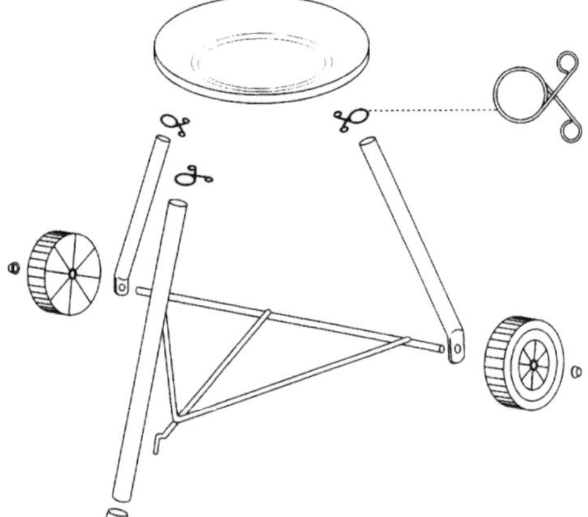

Read important DANGERS, WARNINGS and CAUTIONS in Owner's Guide before operating this barbecue.

30786 11/93

FIGURE 7.2 **Example of Specific Instructions, page 1**

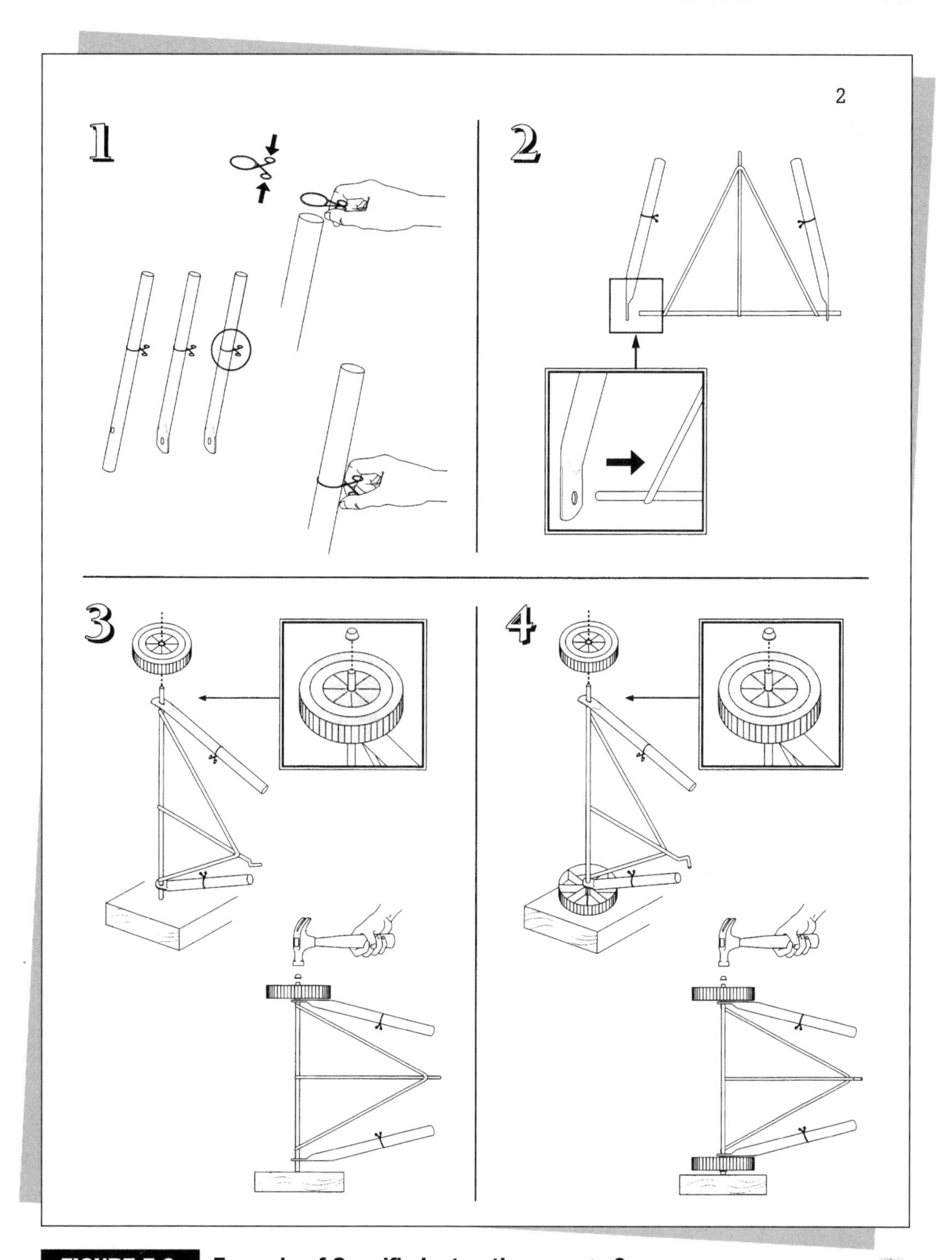

FIGURE 7.2 Example of Specific Instructions, page 2

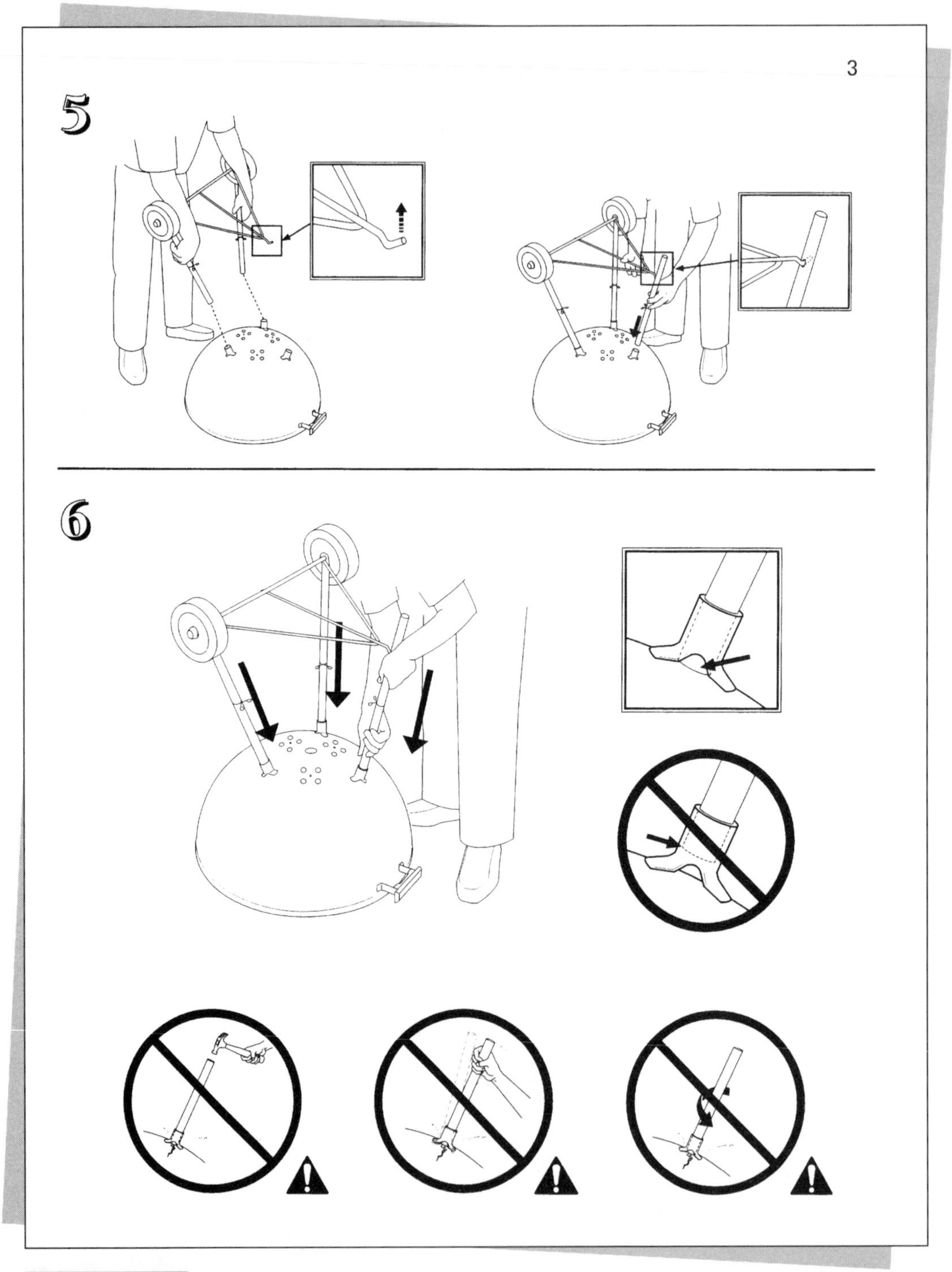

3

5

6

FIGURE 7.2 Example of Specific Instructions, page 3

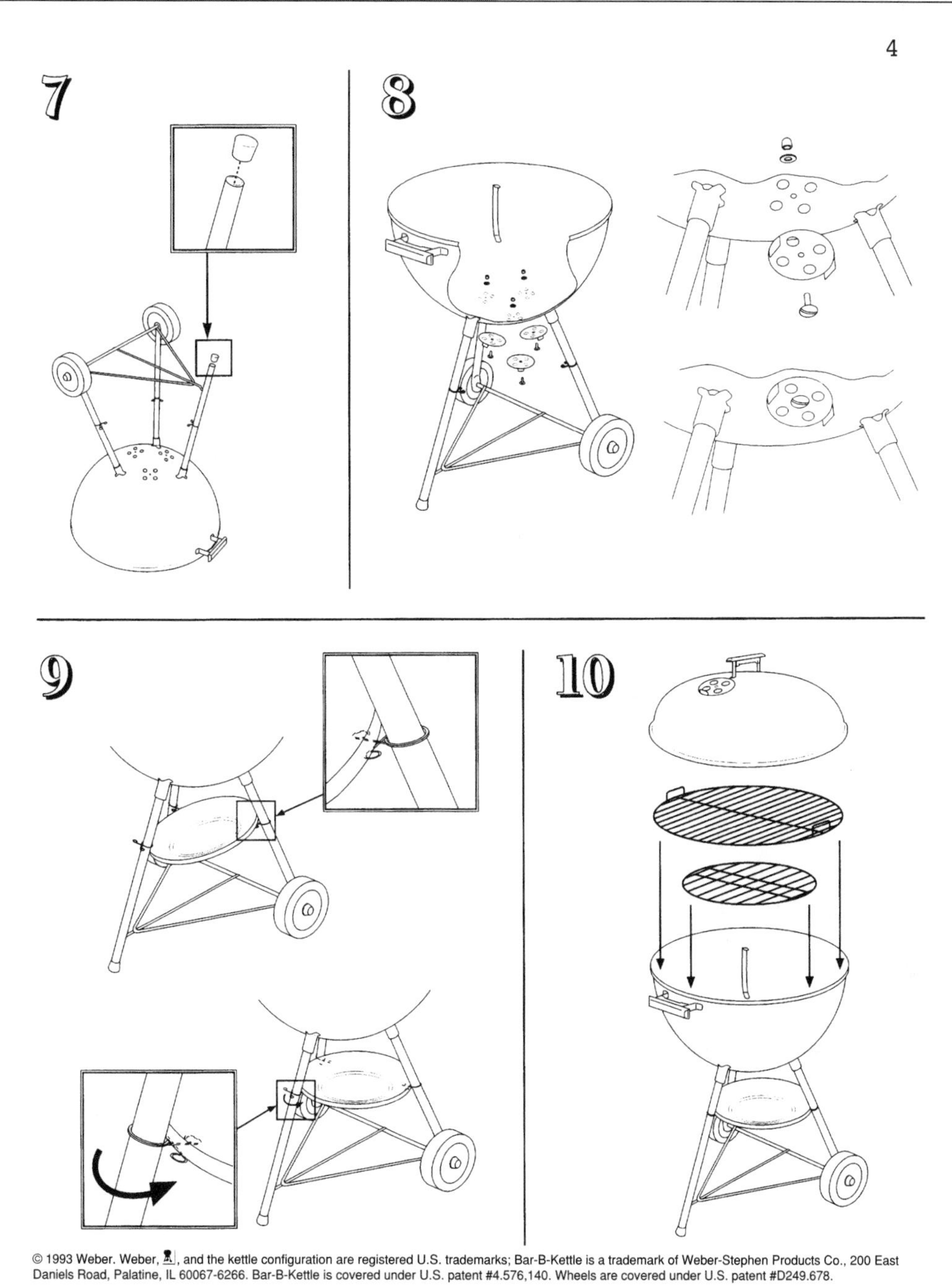

FIGURE 7.2 **Example of Specific Instructions, page 4**

☑ Checklist Evaluating Instructions

A good set of instructions

___ opens with a brief introduction that identifies the procedure and explains its function;

___ lists the materials, equipment, tools, and skills required to perform the procedure;

___ provides a well-organized, step-by-step explanation of how to perform the procedure;

___ provides any appropriate warnings, cautions, or notes to enable the reader to perform the procedure without unnecessary risk;

___ is clear, accurate, and sufficiently detailed to enable the reader to perform the procedure without unnecessary difficulty;

___ employs helpful comparisons and analogies to clarify difficult concepts;

___ uses clear, simple "commands";

___ concludes with a brief summary;

___ employs effective visuals (photographs and line drawings—exploded and/or cutaway views) to clarify the text;

___ contains no typos or mechanical errors in spelling, capitalization, punctuation, and grammar.

Exercises

EXERCISE 7.1

As discussed in the text and mentioned on the checklist, written instructions should open with a brief introduction that identifies the procedure and explains its purpose. Write such introductions for three of the following procedures:

- balancing a checkbook
- creating back-up computer discs
- measuring a person's blood pressure
- programming a digital wristwatch
- hemming a pair of pants

■ EXERCISE 7.2

Write instructions (and include visuals) explaining how to perform a common procedure related to your field of study or employment—for example, welding a lap joint, administering a bed bath, restraining a potentially troublesome person.

■ EXERCISE 7.3

Write instructions (and include visuals) explaining how to perform a common indoor household chore—for example, washing laundry, cleaning a bathroom, repairing a leaky faucet.

■ EXERCISE 7.4

Write instructions (and include visuals) explaining how to perform a common outdoor household chore—for example, mowing a lawn, sealing a blacktop driveway, installing a rain gutter.

■ EXERCISE 7.5

Write instructions (and include visuals) explaining how to perform a common sports-related procedure—for example, waxing a pair of skis, suiting up for an ice hockey game, performing stretching exercises.

■ EXERCISE 7.6

Write instructions (and include visuals) explaining how to perform a common procedure related to automobile maintenance and repair—for example, changing engine oil and filter, using jumper cables, washing and cleaning a vehicle.

■ EXERCISE 7.7

Write instructions (and include a map) explaining how to travel from your home to your college.

■ EXERCISE 7.8

Write instructions (and include visuals) explaining how to perform one of these procedures:

- cooking a favorite meal
- laying track on a model railroad
- building a dog house
- changing a baby's diaper
- waterproofing a pair of work boots

■ EXERCISE 7.9

Consult the owner's manual accompanying a household appliance or other product you have purchased and examine the visuals provided to facilitate assembly or proper use. How helpful are they? Write a memo report to your instructor in which you discuss your findings. Be sure to include copies of the visuals.

■ EXERCISE 7.10

Consult some textbooks, periodicals, and Web sites devoted to your field of study or employment and find five examples of visuals designed to accompany instructions. How helpful are they? Write a booklet report to your instructor, in which you discuss your findings. Be sure to include copies of the visuals.

8

Job Application Process: Letter, Résumé, Interview, and Follow-Up

Learning Objective Upon completing this chapter, you will be able to write an effective job application letter and résumé, and to interview and follow up successfully.

The employment outlook today is quite challenging, with many qualified applicants vying for every available position, and all indicators suggest that the competition will soon get even tougher. Therefore, it is now more important than ever to fully understand the process of applying for a job. Essentially it involves four components: an effective cover letter, an impressive résumé, a strong interview, and a timely follow-up.

Some classified ads provide only a telephone number or an address, with no mention of a written response. In such cases, you will likely complete a standardized questionnaire instead of submitting a letter and résumé. Generally, however, these are not the most attractive positions. Ads for the better openings, those that pay more and offer greater opportunity for advancement, typically instruct you to respond in writing. Selective employers deliberately word their ads this way, to weed out the less desirable applicants—those unable to compose a letter and résumé, and those insufficiently motivated to do so.

It is important to understand that employers require a written response partly to secure a representative sample of your *best work*. Therefore, the physical appearance of your job application correspondence is crucially important. You may be well qualified, but if your letter and résumé are sloppy, crumpled, handwritten, poorly typed, or marred by mechanical errors, they will probably be discarded unread. Unless your keyboarding skills and sense of page design are well developed, you might consider using computer software geared to generating résumés (Microsoft Word or WordPerfect, for example), or else hire a professional typist equipped with such resources to prepare your materials. If your application materials are stored on a disk, they can be easily customized to match each position for which you apply. But you cannot expect software—or a typist—to work from scratch. You must understand the basic principles governing the preparation of job application correspondence. This chapter will provide you with that knowledge, along with information about how to ensure a successful interview.

Application Letter

Prepared in a conventional format and three-part structure, a job application letter is no different from any other business letter (see Chapter 2). It should be neatly typed on 8½-by-11-inch standard-bond white paper and should be framed by ample (1- to 1½-inch) margins. In nearly all cases, the letter should be no longer than one page. The writer's address, the date, the reader's name and address, the salutation, and the complimentary close are handled just as they would be in any other letter, except that all punctuation should be provided. Most employment counselors agree that open punctuation and/or an "all caps" inside address should be avoided in an application letter because some personnel directors dislike these practices.

Ideally, the classified ad will provide both the name and the title of the person to contact, as in the following example.

> **ELECTRICIAN:** Permanent, full-time. Associate's Degree, experience preferred. Good salary, benefits. Cover letter and résumé to: Maria Castro, Director of Human Resources, The Senior Citizens' Homestead, 666 Grand Boulevard, Belford, CT 06100. Equal Opp'ty Employer.

Sometimes, however, the ad will not mention an individual's name, but will provide only a title—"Personnel Manager," for example—or simply the company's name. In such cases, telephone the employer and explain that you are interested in applying for the job and would like to know the name and title of the contact person. Be sure to get the correct spelling, and—unless the name plainly reveals gender—determine whether the individual is a man or a woman. This ensures that your letter will be among the only personalized ones received, thereby creating a more positive first impression.

Unfortunately, some ads reveal almost nothing—not even the name of the company—and simply provide a box number at the newspaper or the post office, like this:

> **ACCOUNTING ASSISTANT:** Computer skills and one year hands-on experience with A/R & A/P required. Reply to Box 23, The Bayonne Times, 500 Broadway, Bayonne, NJ 07002.

In such an instance, set up the inside address in your letter as follows:

> Box 23
> The Bayonne Times
> 500 Broadway
> Bayonne NJ 07002

When there is no way to identify whom you are addressing, use "Dear Employer" as your salutation. This is a bit more original than such unimaginative greetings as the impersonal "To Whom It May Concern," the gender-biased "Dear Sir," or the old-fashioned "Dear Sir or Madam."

Again, your letter will stand out from the others received, suggesting you are more resourceful than the other applicants.

In your opening paragraph, directly state your purpose: that you are applying for the job. Strangely, many applicants fail to do so. Wordy and ultimately pointless statements such as "I read with great interest your classified advertisement in the Tuesday edition of my hometown newspaper, *The Daily Gazette*" invite the reader to respond, "So? Do you *want* the job, or what?" Instead, compose a one-sentence opening that comes right to the point: "As an experienced sales professional, I am applying for the retail position advertised in *The Daily Gazette.*" This straightforward approach suggests that you are a confident, focused individual—and therefore a desirable applicant.

Always mention the job *title,* as the employer may have advertised more than one. Also indicate how you learned of the job opening. Most employers find this information helpful in monitoring the productivity of their various advertising efforts, and they appreciate the courtesy. If you learned of the opening through word of mouth, however, do *not* mention the name of the person who told you about it, even if you have been given permission to do so. The individual may not be well regarded by the employer, and since you have no way of knowing this, you should not risk the possibility of an unfortunate association. In such a situation, use a phrase such as "*It has come to my attention that you have an opening for an electrician,* and I am applying for the job."

In your middle paragraph(s), provide a narrative summary of your experience, education, and other qualifications. Go into some depth, giving sufficient information to make the employer want to read your résumé, which should be specifically mentioned. But avoid *excessive* detail. Dates, addresses, and other such specifics belong in the résumé, not the letter. Be sure to mention, however, any noteworthy attributes—specialized licenses, security clearances, computer skills, foreign language fluency—that may set you apart from the competition. Do not pad the letter with vague claims that cannot be documented. "I have five years of continuous experience as a part-time security guard" scores a lot more points than "I am friendly, cooperative, and dependable." Never mention weaknesses, and always strive for the most upbeat phrasing you can devise. "I'm currently unemployed," for example, creates a negative impression; the more positive "I am available immediately" turns this circumstance to your advantage.

The purpose of this process, of course, is to make the employer recognize your value as a prospective employee. Using the "you" approach explained in Chapter 1, gear your letter accordingly. Without indulging in exaggeration or arrogant self-congratulation, explain why it would be in the employer's own best interests to hire you. Sometimes an honest, straightforward statement such as this can be quite persuasive: "With my college education now completed, I am very eager to begin my career in banking and will bring a high level of enthusiasm and commitment to this position."

Your closing paragraph—no longer than two or three sentences—should briefly thank the employer for considering you and should request an interview. Nobody has ever received a job offer on the strength of a letter alone. The letter leads to the résumé, the résumé (if you are lucky) secures an interview, and the interview (if you are *really* lucky) results in a job offer. By mentioning both the résumé and the interview in your letter, you indicate that you are a knowledgeable person who is familiar with conventions of the hiring process.

Understand, however, that even one mechanical error in your letter may be enough to knock you out of the running. You must make absolutely certain that there are no typos, spelling mistakes, faulty punctuation, or grammatical blunders—none whatsoever! Check and double-check to ensure that your letter (along with your résumé) is mechanically perfect.

Figure 8.1 depicts an effective application letter in response to the classified ad for an electrician, on page 183. The accompanying résumé (Figure 8.2) will enable you to see how the résumé and letter interrelate.

Résumé

As Figure 8.2 illustrates, a résumé is basically a detailed list or outline of a job applicant's work history and other qualifications. The following categories of information typically appear:

- Personal Information
- Career Objective
- Education
- Work Experience
- Military Service
- Specialized Skills or Credentials
- Honors and Awards
- Community Activities

Of course, few résumés include *all* of these categories. Not everyone has served in the military, for example, or received awards. Not everyone is active in the community or possesses special skills. But practically

32 Garfield Avenue
Belford, CT 06100
April 2, 1999

Ms. Maria Castro
Director of Human Resources
The Senior Citizens' Homestead
666 Grand Boulevard
Belford, CT 06100

Dear Ms. Castro:

As an experienced electrician about to graduate from County Community College with an A.O.S. degree in Electrical Engineering Technology, I am applying for the electrician position advertised in *The Daily Herald*.

In college I have maintained a 3.60 grade-point average while serving as Vice-President of the Technology Club and Treasurer of the Minority Students Union. In keeping with my ongoing commitment to community service, last year I joined a group of volunteer workers renovating the Belford Youth Club. Under the supervision of a licensed electrician, I helped rewire the building, and acquired a great deal of practical experience during the course of this project. The combination of my academic training and the hands-on knowledge gained at the Club equips me to become a valued member of your staff. Past and current employers—listed on the enclosed résumé—will attest to my strong work ethic. I can provide those individuals' names and telephone numbers upon request.

Thank you very much for considering my application. Please telephone me to arrange an interview at your convenience.

Sincerely,

James Carter

James Carter

FIGURE 8.1 Application Letter

James Carter

32 Garfield Ave., Belford, CT 06100
(203) 555-2557

Career Objective

To secure a permanent, full-time position as an electrician.

Education

County Community College (1996–present)
1101 Belford Dr., Belford, CT

Will graduate in May 1999 with an A.O.S. degree in electrical engineering technology. Have maintained a 3.60 grade-point average while serving as Vice-President of the Technology Club and Treasurer of the Minority Students' Union.

Experience

Counter Clerk (1996–present)
Quik Stop Grocery, 255 Bergen St., Belford, CT

Part-time position to help meet college expenses.

Warehouse Worker (1993–1996)
S. Lewis & Sons, 13 North Rd., Belford, CT

Worked this job after high school before deciding to pursue college education.

Community Activities

Assistant Little League baseball coach; member, church choir; volunteer, Belford Youth Club renovation project (helped rewire building).

FIGURE 8.2 **Résumé**

anyone can assemble an effective résumé. The trick is to carefully evaluate your own background, identify your principal strengths, and emphasize those attributes. A person with a college degree but with little relevant experience, for example, would highlight the education component. Conversely, someone with a great deal of experience but relatively little formal schooling would instead emphasize the employment history. Both individuals, however, would follow these well-established guidelines that govern all résumés:

1. The résumé, like the cover letter, should be visually attractive. It should be typed or laser-printed on 8½-by-11-inch standard-bond white paper only. Use capitalization, boldface, and white space skillfully to create an inviting yet professional appearance. Unlike a letter, a résumé can be laid out in any number of different ways, but it must *look good.* Experiment with a variety of layouts until it does.

2. The various categories of information must be clearly labeled, and must be distinct from one another. This enables the employer to quickly review your background without having to labor over the page. Indeed, most employers are unwilling to struggle with a confusing résumé, and will simply toss it aside and move on to the next one.

3. All necessary details must appear—names, addresses, dates, etc.—and must be presented in a consistent manner throughout. For example, do not abbreviate words like "Avenue" and "Street" in one section and then spell them out elsewhere. Adopt one approach to abbreviation, capitalization, spacing, and other such matters.

4. Use *reverse*-chronological order in sections such as "Education" and "Work Experience." List the most recent information first, then work backwards through time.

5. A résumé should not be longer than one page unless the applicant's background and qualifications truly warrant a second. This is not usually the case, except among applicants with ten or more years of workplace experience. If your résumé does have a second page, include your name at the top.

6. Make sure the résumé is mechanically perfect, with absolutely no errors in spelling, punctuation, or grammar. Edit for careless blunders—typos, inconsistent spacing, and the like.

Here are some detailed pointers concerning the specific categories of information:

- ***Personal Information.*** Such irrelevant details as birth date, religion, marital status, social security number, and so forth simply waste space. Include *only* your name, address, and phone number. If you have a fax number or e-mail address, include these too. This section should appear at the top of the page, and need not be labeled.
- ***Career Objective.*** A brief but specific statement of your career plans, this can help if you intend to apply for only one kind of job. But if you wind up applying for a wider range of possibilities, it can sabotage your chances unless you are willing to revise it to suit each occasion. Whether to include this category depends on your individual circumstances.
- ***Education.*** In *reverse*-chronological order, provide the name and address of each school you attended and mention your program of study and any degrees, diplomas, or certificates received, along with dates of attendance. You may wish to list specific classes completed, but this consumes a lot of space and is not necessary. Do not list any schooling earlier than high school, and high school itself should be omitted unless you are attempting to "beef up" an otherwise skimpy résumé.
- ***Work Experience.*** Along with the Education section, this is the most important part of the résumé. For each position you have held, provide your job title, dates of employment, the name and address of your employer, and—if they are not evident from the job title—the duties involved. Some résumés also include the names of immediate supervisors. As in the Education section, use *reverse*-chronological order. If you have worked at many different jobs, some for short periods of time, you may list only your most important positions, omitting the others or lumping them together in a one-sentence summary, like this: "At various times I have also held temporary and part-time positions as service station attendant, counter clerk, and maintenance worker."
- ***Military Service.*** If applicable, list the branch and dates of your service, the highest rank you achieved, and any noteworthy travel or duty. Some applicants, especially those with no other significant employment history, list military activity under the Work Experience category.

- ***Specialized Skills or Credentials.*** Include licenses, certifications, security clearances, foreign language competency, proficiency with certain machines—any "plus" that does not fit neatly elsewhere.

- ***Honors and Awards.*** These can be academic or otherwise. In some cases—if you received a medal while in the military, for example, or made the college honor roll—it is best to include such distinctions under the appropriate categories. But if the Kiwanis Club awarded you its annual scholarship or you were cited for heroism by the mayor, these honors would probably be highlighted in a separate section.

- ***Community Activities.*** Volunteer work or memberships in local clubs, organizations, or church groups are appropriate here. Most helpful are well-known activities such as Scouting, Little League, 4-H, PTA, and the like. Include full details: dates of service or membership, offices held, if any, and special projects or undertakings you initiated or coordinated. Obviously, community activities often bear some relationship to applicants' pastimes or hobbies. Employers are somewhat interested in this, because they're seeking individuals who can not only perform the duties of the job but also "fit in" easily with co-workers. But do not claim familiarity with an organization or activity you actually know little about. You are likely to get caught, since many interviewers like to open with some preliminary conversation about an applicant's interests outside the workplace.

- ***References.*** It is no longer required to list the names of references on a résumé, or even to include a "References available upon request" line. Nevertheless, you will probably be asked for references if you become a finalist for a position, so you should mention near the end of your application letter that you are ready to provide them. But never identify someone as a reference without the person's permission. Before beginning your job search, identify at least three individuals qualified to write recommendation letters for you, and ask them whether they would be willing to do so. Select persons who are familiar with your work habits and who are likely to comment favorably. Teachers and former supervisors are usually the best choices for recommendations because their remarks tend to be taken the most seriously by employers. You must be absolutely certain, however, that anyone writing on your behalf will have nothing but good things to say. Tentative, half-hearted praise is worse than none at all. If someone seems even slightly hesitant to serve as a reference, you should find somebody more agreeable. One way to determine whether someone is indeed willing to compose an enthusiastic endorsement is to request that a copy of the recommendation be sent to you as well as to the employer. Anyone reluctant to comply with such a request is probably not entirely supportive. In any case, securing copies of recommendation letters enables you to judge for yourself whether any of your references should be dropped from your list. Better to suffer the consequences of a lukewarm recommendation once than to be undermined repeatedly without your knowledge. Usually, however, anyone consenting to write a letter on your behalf (and provide you with a copy) will give an affirmative evaluation that will work to your advantage.

Traditional Résumés

A traditional résumé can be organized in accordance with any one of three basic styles: chronological, functional, or a combination of both.

A **chronological résumé** is the most common, and the easiest to prepare. Figure 8.2, the James Carter résumé, typifies this style. Schooling and work experience are presented in reverse-chronological order, with schools' and employers' names and addresses indicated, along with the dates of the applicant's attendance or employment. Descriptions of the applicant's specific job responsibilities or courses of study are provided as part of the Experience or Education categories. This style is most appropriate for persons whose education and past experience are fairly consistent with their future career plans, or for those seeking to advance within their own workplace.

A **functional résumé,** on the other hand, highlights *what* the applicant has done, rather than where or when it has been done. The functional résumé is skills-based, summarizing in general terms the applicant's experience and potential for adapting to new work challenges. Specific chronological details of the person's background are included, but are not the main focus of such a résumé. This style is most appropriate for applicants wishing to emphasize their actual proficiencies rather than their work history.

As the term suggests, the **combination résumé** is a blend of the chronological and functional approaches, featuring a relatively brief "skills" section at the outset, followed by a chronological detailing of the applicant's background. The combination approach is most appropriate for applicants whose experience is relatively diversified, and whose skills span a range of functional areas. Figures 8.3–8.5 depict the same applicant's résumé prepared in each of the three styles.

Carole A. Greco

61 Stebbins Drive
Smallville, NY 13323
(315) 555-5555

OBJECTIVE: A permanent position in financial services.

EDUCATION: Associate in Applied Science (Accounting), May 1999
County Community College, Elliston, NY

GPA 3.65; Phi Theta Kappa Honor Society; Phi Beta Lambda Business Club; Ski Club.

EXPERIENCE: Intern (Fall 1998)
Sterling Insurance Company, Elliston, NY

Contacted and met with prospective clients, answered client inquiries, performed general office duties.

Trust Administrative Assistant (Summers 1996–1998)
First City Bank, Elliston, NY

Researched financial investment data, organized trust account information, screened and answered customer inquiries, composed business correspondence.

Student Congress Treasurer (Fall 1997–Spring 1998)
County Community College, Elliston, NY

Maintained $300,000 budget funding 35 campus organizations, approved and verified all disbursements, administered Student Congress payroll.

SERVICE: Volunteer of the Year, 1998
American Red Cross, Elliston, NY

FIGURE 8.3 Chronological Résumé

Carole A. Greco

61 Stebbins Drive
Smallville, NY 13323
(315) 555-5555

OBJECTIVE

A permanent position in financial services.

COMPETENCIES

Financial
- Interpreted financial investment data
- Assisted with disbursement of trust accounts
- Administered $300,000 budget

Leadership and Management
- Participated in policy-making
- Addressed client/customer concerns

Research and Organization
- Researched financial investment data
- Organized trust account data
- Coordinated funding for 35 organizations

EDUCATION

Associate in Applied Science (Accounting), May 1999, County Community College, Elliston, NY; GPA 3.65; Phi Theta Kappa Honor Society, Phi Beta Lambda Business Club, Ski Club

EXPERIENCE

Intern (Fall 1998), Sterling Insurance Company, Elliston, NY

Administrative Assistant (Summers 1996–1998), First City Bank, Elliston, NY

Student Congress Treasurer (Fall 1997–Spring 1998), County Community College, Elliston, NY

FIGURE 8.4 Functional Résumé

Carole A. Greco

61 Stebbins Drive
Smallville, NY 13323
(315) 555-5555

OBJECTIVE

A permanent position in financial services.

COMPETENCIES

- Strong account management and financial analysis skills
- Effective leadership and management capabilities
- Well-developed research and organizational skills

EXPERIENCE

Intern (Fall 1998), Sterling Insurance Company, Elliston, NY: Contacted and met with prospective clients, answered client inquiries, performed general office duties.

Trust Administrative Assistant (Summers 1996–1998), First City Bank, Elliston, NY: Researched financial investment data, organized trust account information, screened and answered customer inquiries, composed business correspondence.

Student Congress Treasurer (Fall 1997–Spring 1998), County Community College, Elliston, NY: Maintained $300,000 budget funding 35 campus organizations, approved and verified all disbursements, administered Student Congress payroll.

EDUCATION

Associate in Applied Science (Accounting), May 1999, County Community College, Elliston, NY; GPA 3.65; Phi Theta Kappa Honor Society, Phi Beta Lambda Business Club, Ski Club

SERVICE

Volunteer of the Year (1998), American Red Cross, Elliston, NY

FIGURE 8.5 Combination Résumé

Scannable Résumés

Rapid advances in computer technology have greatly changed every aspect of workplace communications. The hiring process is no exception. Many companies now use electronic scanners to load letters and résumés into databases containing all applications received, enabling personnel directors to evaluate applicants' credentials on the computer.

One problem with this development is that, depending on which software is used, a creatively formatted résumé may appear confusingly jumbled on the screen. Although some systems will preserve the résumé's original appearance, others simply read from left to right, without necessarily recognizing headings and columns as such, or even observing the breaks between lines. Moreover, design features such as alternative typefaces, underlining, and italics may play havoc with the electronic scanner's recognition capabilities, producing practically illegible results. For this reason, many career counselors now urge applicants to greatly simplify the design of their résumés by adopting a no-frills, flush-left format less likely to create difficulties of this nature. Figure 8.6 depicts the same résumé as in Figure 8.3, but this time in a scannable format.

In the final analysis, simplification for the computer's sake is probably a positive development. Simpler is generally better in workplace communications, and the trend toward a more streamlined résumé layout serves to counterbalance the tendency to overdesign such documents. The simpler formats also permit more information to be included. Notice, for example, that although the traditional résumé in Figure 8.3 and the scannable version in Figure 8.6 contain the same information, the scannable résumé uses less of the page, creating the option of slightly expanding the scope of the résumé if desired.

In a related trend, it is now possible to explore the Internet for job postings and other such information (searching by company name, job title, geographical location, or other key word phrase), and even to post your own credentials to on-line résumé banks consulted by employers worldwide. Here are several such sites:

Career Mosaic <http://www.careermosaic.com>
NationJob <http://www.nationjob.com>
Net Temps <http://www.net-temps.com>

These sites are accessible through links listed in services such as The Career Search Launch Pad at <http://www.anet-dfw.com/~tsull/career/cslp.html> or can be reached directly, by using Net browsers such as

Carole A. Greco

61 Stebbins Drive
Smallville, NY 13323
(315) 555-5555

OBJECTIVE: A permanent position in financial services.

EDUCATION: Associate in Applied Science (Accounting), May 1999, County Community College, Elliston, NY: GPA 3.65; Phi Theta Kappa Honor Society, Phi Beta Lambda Business Club, Ski Club

EXPERIENCE: Intern (Fall 1998), Sterling Insurance Company, Elliston, NY: Contacted and met with prospective clients, answered client inquiries, performed general office duties.

Trust Administrative Assistant (Summers 1996–1998), First City Bank, Elliston, NY: Researched financial investment data, organized trust account information, screened and answered customer inquiries, composed business correspondence.

Student Congress Treasurer (Fall 1997–Spring 1998), County Community College, Elliston, NY: Maintained $300,000 budget funding 35 campus organizations, approved and verified all disbursements, administered Student Congress payroll.

SERVICE: Volunteer of the Year (1998), American Red Cross, Elliston, NY.

FIGURE 8.6 **Scannable Résumé**

Microsoft Explorer or Netscape Communicator, which enable you to seek information by using the uniform resource locators (URL's) that begin with <http. After gaining access to the online site, you simply select menu options to locate the information and services desired: company profiles, résumé-writing tips, on-line classifieds, and so on. Obviously, these resources can be quite helpful to a job-seeker, and are continuously expanding.

Interview

If your letter and résumé result in an interview, the employer considers you at least reasonably well qualified for the job. No personnel office deliberately wastes time interviewing applicants who are not. You can assume, therefore, that you are in the running for the position. But you must outperform the other finalists. To succeed in the employment interview, you must have three assets going for you: preparation, composure, and common sense.

To prepare, find out everything you can about the position and the workplace. Read any existing literature about the employer (annual reports, promotional materials, product brochures, and so on). Consult some of the employment-related Web sites mentioned earlier in this chapter. If possible, talk to employees at the company in question or in comparable jobs elsewhere. For generic information about the job title, consult the United States Department of Labor's *Occupational Outlook Handbook*. (A librarian can direct you to this and similar print and on-line resources.) By familiarizing yourself with the nature of the job and the work environment, you'll better equip yourself to converse intelligently with the interviewer. You'll *feel* more confident, a major prerequisite to interviewing successfully.

Be sure to get enough sleep the night before the interview. Take a shower. Eat breakfast. Dismiss from your mind all problems or worries. If possible, locate the interview site the day before and determine how much time you will need to get there punctually. All this may seem like obvious and rather old-fashioned advice, but it goes a long way toward ensuring that you will be physically and mentally at ease, ready to interact smoothly. Of course, you should not be *too* relaxed; an employment interview is a fairly formal situation and you should conduct yourself accordingly. Stand up straight, shake the interviewer's hand firmly, establish eye contact, and speak in a calm, clear voice. Sit down only when invited to, or when the interviewer does. Do not smoke or chew gum, and—needless to say—NEVER attend an employment interview with alcohol on your breath or while under the influence of any controlled substance.

Anticipate key questions. As mentioned earlier, the interviewer is likely to begin by asking you something about your interests and hobbies, or perhaps by remarking about the weather or some other commonplace matter. From there the conversation will probably become more focused. Expect discussion of your qualifications, your willingness to work certain hours or shifts, your long-range career plans, your desired salary, your own questions about the job, and so on. Here is a list of twenty typical questions often asked by interviewers:

1. Tell me about yourself. (A common variation is, If you had to describe yourself in just one word, what would it be?)
2. What do you do in your spare time?
3. Why did you choose your particular field of study?
4. What do you think you've learned in college?
5. How much do you know about computers?
6. Why aren't your grades higher?
7. Do you plan to further your education?
8. What are your long-range goals?
9. What kind of work do you like best? Least?
10. What was the best job you ever had? Why?
11. How do you explain the "gaps" on your résumé?
12. Can you provide three solid references? Who?
13. Why do you think you're qualified for this position?
14. Why do you want to work for this particular company?
15. Are you willing to work shifts? Weekends? Overtime?
16. Are you willing to relocate?
17. If hired, when could you start work?
18. How much do you expect us to pay you?
19. Why should we hire you?
20. Do *you* have any questions?

Interviewers ask questions such as these partly because they want to hear your answers, but also because they want to determine how poised you are, how clearly you express yourself, and how well you perform under pressure. Try to formulate some responses to such queries beforehand so you can respond readily, without having to search your mind. Just as importantly, try to settle upon several good questions of your own. Gear these to matters of importance, such as the employer's training and orientation procedures, the job description and conditions of employment, performance evaluation policies, likelihood of job stability, opportunities for advancement, and the like. You want the employer to know that you are a serious candidate with a genuine interest in the job. But do not talk *too* much or attempt to control the interview. Answer questions fully—in three or four sentences—but know when to stop talking. Stay away from jokes or controversial topics. Avoid excessive slang. Do not try to impress the interviewer with "big words" or exaggerated claims. Maintain a natural but respectful manner. In short, just be yourself. But be your *best* self.

Many applicants are unsure about how to dress for an interview. The rule is actually quite simple: Wear approximately what you would if you were reporting for work. If you are applying for a "dress-up" job, dress up. For a "jeans and sweatshirt" job, dress casually. Some employment counselors advise applicants to dress just a step above the position for which they are interviewing. In any case, make sure your interview clothing fits properly, is neat and clean, and is not too outlandishly "stylish." Minimize jewelry. Earrings are acceptable (two or more for a woman, one for a man), but nose and eyebrow rings, tongue studs, chains, or any other such adornments are better left at home, as are sweatpants, hats, and any clothing imprinted with crude or tasteless slogans. Just as the physical appearance of your letter and résumé will influence whether you are invited to an interview, your *own* appearance will influence whether you get hired. As mentioned at the beginning of this section, *common sense* is a major factor in interviewing well. You must "use your head" and "put your best foot forward." As threadbare as these well-known clichés may seem, they really are good advice.

Follow-Up

The follow-up to an interview is another example of common sense. Although it requires very little effort, many applicants neglect it. This is unfortunate for them, because a timely follow-up (a few days after the interview) can serve as a tie-breaker among several comparably qualified candidates. Every employer wants to hire someone willing to go a bit beyond what is required. Your follow-up is evidence that you are such a person, and it can therefore enable you to get a step ahead of the other applicants.

Simply a "thank you" letter, the follow-up expresses gratitude for the interview and assures the employer that you are still interested in the job. There is no need to compose anything elaborate; a brief note will do. For an example, see Figure 8.7.

32 Garfield Avenue
Belford, CT 06100
April 15, 1999

Ms. Maria Castro
Director of Human Resources
The Senior Citizens' Homestead
666 Grand Boulevard
Belford, CT 06100

Dear Ms. Castro:

Thank you for meeting with me last week to discuss the electrician position.

Having enjoyed our conversation and the tour of the Senior Citizens' Homestead, I am still very interested in the job, and am available to start work immediately after my graduation from college next month. I can also start sooner (on a part-time basis) if necessary.

Please contact me (555-2557) if you have any further questions about my background or credentials, and thanks again for your time.

Sincerely,

James Carter

James Carter

FIGURE 8.7 Follow-up Letter

Evaluating an Application Letter, Résumé, and Follow-Up

A good application letter

___ follows standard format;

___ breaks down into paragraphs:

- ☐ first paragraph asks for the job, by name, and indicates how you learned of the opening,
- ☐ middle paragraphs briefly outline your credentials and refer the reader to your résumé,
- ☐ last paragraph closes on a polite note, mentioning that you would like an interview;

___ does not exceed one page;

___ uses simple language, maintains appropriate tone, and contains no typos or mechanical errors in spelling, capitalization, punctuation, and grammar.

A good résumé

___ looks good, making effective use of white space, capitalization, boldface, and other format features;

___ includes no irrelevant personal information;

___ includes separate, labeled sections for education, experience, and other major categories of professional qualifications;

___ maintains a consistent approach to abbreviation, spacing, and other elements;

___ does not exceed one page;

___ contains no typos or mechanical errors in spelling, capitalization, punctuation, and grammar.

A good follow-up letter

___ follows standard format;

___ breaks down into paragraphs:

- ☐ first paragraph thanks the employer for the interview and mentions the job by name,
- ☐ middle paragraph re-states the applicant's interest and availability,
- ☐ last paragraph politely invites further contact;

___ does not exceed one page;

___ uses simple language, maintains appropriate tone, and contains no typos or mechanical errors in spelling, capitalization, punctuation, and grammar.

Exercises

EXERCISE 8.1

Read the classified advertisements in a recent issue of your local newspaper and write a booklet report about what you find there. Include information not only about what kinds of jobs are listed, but also about the qualifications required. Provide a breakdown of how many jobs require written responses as opposed to telephone or personal contact. Indicate whether there appears to be any correlation between the type of job and the likelihood that a written response will be requested.

EXERCISE 8.2

Using the URLs listed on page 195, explore the employment-related Web sites Career Mosaic, NationJob, and Net Temps and write a memo report about what you discover. Compare and evaluate these sites. Which one is best for your purposes? Why? Which is the *least* useful to you? Why?

EXERCISE 8.3

Using the Internet along with print resources such as the Department of Labor's *Job Information Handbook,* research a particular job title and write a booklet report discussing your findings. What are the principal responsibilities of the position? What qualifications are typically required? What is the salary range? Are such jobs more plentiful in certain geographical areas?

EXERCISE 8.4

Using the Internet along with print resources such as Standard and Poor's, research a particular employer and write a booklet report discussing your findings. What are the employer's main products or services? How long has the employer been in business? Where is the corporate headquarters? How large is the workforce? What kinds of skills or credentials are required to work for this company?

■ EXERCISE 8.5

Interview someone currently employed in a job related to your field of study and write a memo report summarizing the conversation. Why did the person choose this kind of work? How long has the person been in the position? What kind of education and other qualifications does the individual possess? What are the best and worst features of the job? Does he or she find the work challenging, interesting, and rewarding?

■ EXERCISE 8.6

Find an actual classified advertisement for an opening in your field that specifically requests a written response. Compose a job application letter and a chronological résumé. Pretend you have been successful in getting an interview, and have met with the personnel director. Compose a follow-up letter.

■ EXERCISE 8.7

Find an actual classified advertisement for an opening in some field unrelated to your own that specifically requests a written response. Compose a job application letter and a functional résumé. Pretend you have been successful in getting an interview, and have met with the personnel director. Compose a follow-up letter.

■ EXERCISE 8.8

Find an actual classified advertisement for an opening in some field other than your own but related to it, which specifically requests a written response. Compose a job application letter and a combination résumé. Pretend you have been successful in getting an interview, and have met with the personnel director. Compose a follow-up letter.

■ EXERCISE 8.9

Design scannable versions of the chronological, functional, and combination résumés created in response to Exercises 8.6–8.8.

EXERCISE 8.10

Three application letters accompanied by résumés follow. For a variety of reasons, all are badly flawed. Rewrite each to eliminate its particular weaknesses.

Carla Zogby
2400 Front St., Apt. 32
Kansas City, MO 64100

February 23, 1999

Diversified Services, Inc.
500 Tower Street
Kansas City, Missouri 64100

Dear Sirs:

I am writting this letter in reply to your recent add in the *Kansas City Star.*

As you can see from the enclosed resume, I have all the qualifications for which you are looking for.

Thank you for your time.

Your's Truely,

Carla Zogby

Carla Zogby

EXERCISE 8.10 Continued

NAME: Carla Zogby
ADDRESS: 2400 Front St., Apt. 32, Kansas City, MO 64100
TELEPHONE MUMBER: 816-555-4370
DATE OF BIRTH: October 1, 1975
RELIGION: Cathoilc
MARITAL STATUS: Single
HEIGHT: 5"3' WEIGHT: 110 lbs.

EXPERIENCE

9/94–2/95 Receptionist	St. Aedan's Church Answered phones, greeted visitors, handled weekly collection deposits, prepared and distributed weekly bulletin.
3/95–8/96 Store Trainer, Waitress	Friendly's Corporation Trained all new waitstaff, took food orders, cleared tables, washed dishes, helped cook.
11/96–present Insurance Processor	City Bank Process disability and death claims, work with insurance companies to pay accounts.

EDUCATION

9/90–6/94	St. Aedan's High School • Honor Roll 3, 4 • Student Council 2
8/98–present	Kansas City Technical College Secretarial Science

SKILLS
Personal computer systems, software proficiency with spreadsheets, word processing and database programs.

EXERCISE 8.10 Continued

Thomas Logan
105 Lincoln Ave.
Lincoln, Nebraska 68500
July 17, 1999

Conklin's Department Stores, Inc.
1400 West Carroll Street
Chicago, Illinois 60600

Gentlemen,

I am responding to an employment ad of yours that I found via the Internet for the Store Security postion. I am sure that you will find that I am highly qualified for this job.

As a military policeman in the United States Army from July 1993 until April 1999 I had over two years experience in law enforcement. My job responsibilities included public relations, emergency vehicle operations, weapons handling, equipment maintenance and personnel management. My training included interpersonal communication skills, radio communications procedures, weapons safety, police radar operations, unarmed self-defense and riot and crowd control operations. I enforced traffic regulations by monitoring high traffic areas, being visible to the public, and issuing citations as necessary. I performed law enforcement investigations as needed, as well as prepared, verified, and documented police reports to include sworn statements and gathering and processing evidence. I conducted foot and motorized patrols of assigned areas and applied crime prevention measures by maintaining control and discipline through ensuring that all laws and regulations were obeyed at all times. I also performed basic First Aid as first responder when needed.

Earlier I served as a parachute rigger, rigging, assembling, and repairing several of the military parachutes used in Airborne operations. I rigged various vehcles, weapons, and supplies to be air-dropped as well as hold airborne status for the duration. I also trained in combat operations.

At present I have just enrolled in the Criminal Justice program at Lincoln (Nebraska) Community College, and I have also completed a Human Relations course at Texas Central College, a Combat Lifesaver course, a ten-week course at the United States Army Military Police School, as well as studies at the United States Army Airborne and Pararigger Schools and the United States Army Basic Training and Infantry Schools.

Additional Skills include knowledge of First Aid, knowege of conversational Spanish, an accident-free driving record (nine years (civilian and military), and a United States Army Secret Security Clearance.

Sincrely,

Thomas Logan

EXERCISE 8.10 Continued

Résumé of

Thomas Logan, Jr.

105 Lincoln Avenue
Lincoln, Nebraska

Career Objective
Full-time position in law enforcement or security.

Education
Dickinson High School
Jersey City, NJ (Class of 1993)

Lincoln Community College
Lincoln, Nebraska (Currently enrolled)

Armed Forces
United States Army (1993—1999)

Interests
Fishing, Hunting, Snowmobiling

References
Professor John Dhayer
Sgt. Warren Landis
Mr. Thomas Logan, Sr.

■ EXERCISE 8.10 Continued

July 17, 1999

Superior Steel, Inc.
c/o NYS Department of Labor
121 North Main Street
Herkimer, NY 13350

Dear Superior Stell;

I am applying for the machinist/production assembler position you have posted with the Depratment of Labor. I have been a machinist at the Curtis Arms Co. for two years with experience in the manufacture of low tolerance parts from blueprints. I also have eight years experience as a self-employed general contractor, and additional experience as a tree service worker. I am now continuing my education at Proctor Technical College. I have completed 12 credits towArd an AOS degree and have maintained a 4.0 GPA. I am looking forward to meeting with you for an interview as soon as possible. Thank you for your consideration.

Sincerely;

Roland Perry

Roland Perry

■ EXERCISE 8.10 Continued

Roland Perry
3 East Street
Proctor, NY 13500
(315)555-3806

OBJECTIVE

To obtain a full-time position as a machinist with Superior Steel.

WORK HISTORY

Curtis Arms Company, Inc., Utica, NY

Machine operation/set-up on CNC, Pratt-Whitney bore reamers, NAPCO black oxide color line, neutral and hardening furnaces. (January 1997–present)

DUTIES

- machining gun parts to tolerances of +/– .005 inch
- metal fabricating from blue prints
- hardening parts to Rockwell hardness specifications
- maintaining quality standards

Larry's Tree Service, Kingston, NY

Ground crew member. (July 1996–January 1997)

DUTIES

- operated chain saws, chippers, stump machine
- controlled lowering lines and climbers lifeline
- operated and maintained trucks and machinery

EXERCISE 8.10 Continued

-page 2-

Perry Construction, Inc., Poughkeepsie, NY

Self-employed general contractor. (August 1986–February 1996)

DUTIES

- carpentry, masonary, plumbing, electrical work
- contracts, book-keeping, customer service

EDUCATION

Proctor Technical College, Proctor, NY (August 1998–present)

COURSES COMPLETED

- Air Conditioning Technology 101–A
- Technical English 101–A
- Technical Math 101–A
- Public Speaking 101–A

COMMUNITY ACTIVITIES

Volunteer Fire Department
American Legion Post Secretary
Community Band (Tuba Player)

9

Oral Presentations: Preparation and Delivery

Upon completing this chapter, you will be able to prepare and deliver successful oral presentations.

Most people dread the prospect of having to stand in front of an audience and make a speech. They feel unsure of themselves, and fear they will appear awkward or foolish. Nevertheless, you should make a real effort to overcome such misgivings. The ability to present your ideas clearly and forcefully to a group of listeners is a valuable skill that equips you for leadership in the workplace, where it is often necessary to address groups of supervisors, co-workers, clients, or customers. It is also quite useful in community contexts such as club gatherings, town meetings, school board hearings, and other public forums. It is certainly helpful in the college setting too, where oral reports are becoming a requirement in more and more courses.

A good speech is the result of three elements: preparation, composure, and common sense. If that sounds familiar, it should; in Chapter 8 the same was said of the employment interview. In many respects the two endeavors are similar. Both are examples of oral communication, both are fairly formal speaking situations, and both place essentially the same demands on you. The main difference, of course, is that in a job interview you are usually speaking to one or two listeners, whereas an oral presentation generally involves addressing a group. This chapter will teach you to prepare and deliver successful oral presentations.

Preparation

A successful oral presentation nearly always is based on thorough preparation beforehand. Preparation involves some preliminary activities we will now consider, followed by actual rehearsal of the speech.

Preliminaries

Preparing for an oral presentation is much like preparing to write. Just as if you were about to compose a memo, letter, or written report, you must first identify your purpose. Are you simply trying to inform your listeners, or are you attempting to entertain them? Are you perhaps seeking to persuade them of something or motivate them to action? In any case, you need a plan that will enable you to achieve your goal.

It is crucial to assess your audience. What are your listeners' backgrounds and interests? How about their perspective on your topic? In short, what might influence their expectations or responses? Unless you gear your remarks to your audience, you will probably not connect satisfactorily with your listeners. For example, a mayoral candidate addressing a gathering of senior citizens would be foolish to focus a campaign speech on long-range outcomes the listeners may never live to see. Such a group would respond better to a presentation of the candidate's short-term goals, particularly those related to that audience's immediate concerns—crime prevention, perhaps, or health care. Just as you do in written communications, you must always bear in mind the nature of your audience when preparing your remarks.

It is also helpful to get a look in advance at the room where you will be speaking. This ensures that you will be somewhat more at ease during the presentation, because you will be on familiar turf. If you are planning to use audiovisual equipment, you should acquaint yourself with it as well. Nothing is more embarrassing than launching a speech and suddenly discovering that there is no convenient electrical outlet for your slide projector, or that the bulb in the transparency projector is burned out. Guard against such setbacks by checking the equipment when you visit the site beforehand.

And of course you must be thoroughly familiar with your subject matter. Gather information about the topic, assembling an arsenal of facts, figures, and examples to support your statements. This will require some research and homework—an essential part of your preparation. You must know not only how to approach and organize the material, but also how to *develop* it. Nobody wants to listen to a speaker who has nothing to say, or who rambles on and on with no apparent direction or focus.

Therefore, the opening of your speech must include a clear statement of purpose, informing the audience about what to expect. From there you must follow a logical path, covering your material in coherent, step-by-step fashion, dealing with one main idea at a time, in orderly sequence. And you should provide effective transitions to facilitate progress from point to point. For all this to happen, you must write out your entire speech ahead of time. Since it is best, however, to actually *deliver* the speech from notes or note cards, a finely polished, letter-perfect piece of writing is not absolutely necessary. But you do need to have a well-developed and well-organized draft from which notes or cards can be derived by selecting key points and supporting details for each. You must also ensure that your notes or cards are plainly legible, so you can glance down and easily see them on the lectern as you deliver the speech. Prepare your notes or cards using a bold, felt-tipped pen, and write substantially larger than you normally do. It is very damaging to your presentation if you have to pause to decipher your own handwriting, or if you have to bend over or pick up your notes or cards to see them clearly. Figure 9.1 depicts a page from the draft of an oral presentation about various applications of radar technology. Figure 9.2 depicts notes based on that same information, and Figure 9.3 depicts note cards.

As we have seen, radar has obvious military value, and has been used to detect and track enemy planes, submarines, missiles, and so on. Permanent Ballistic Missile Early Warning Systems (BMEWS) are in place at various strategic locations around the globe: Clear, Alaska; Thule, Greenland; Fylingdale Moor, England; and elsewhere. An impressive recent development is Relocatable Over-the-Horizon Radar (ROTHR), which can "bounce high-frequency signals in the 5-28 MHz range off the ionosphere to scan an area from 500 to 1,800 nautical miles away" (Hughes 69). But radar has many non-military applications as well.

Radar permits astronomers to measure interplanetary distances precisely and to collect much data that otherwise might be unavailable, by "obtaining radar echoes from the major bodies of the solar system . . . and deriving as much information as possible from them" (Pettengill 1). Since radar can ascertain surface textures and details and can "find objects as small as insects or as large as mountains" (Thome 64), it's obviously very useful in making maps of distant, restricted, or otherwise inaccessible places—even planets.

Obviously, radar can be nearly as useful to civilian aviators as to the military, by detecting storms and other aircraft, and by determining location and altitude. Indeed, one of the first applications of radar was in radio altimeters. And, of course, air traffic controllers use radar extensively to prevent "runway incursions" and other mishaps. (Nordwall)

FIGURE 9.1 Draft Page of an Oral Presentation

Permanent Ballistic Missile Early
Warning Systems (BMEWS):
Clear, Alaska
Thule, Greenland
Fylingdale Moor, England

Relocatable Over-the-Horizon Radar (ROTHR): bounces high-frequency signals off ionosphere; can scan areas 500-1800 nautical miles away.

Astronomers: measure interplanetary distances, collect solar system data.

Cartographers: make maps—even of planets; can "find objs as small as insects or as large as mts."

Aviators: detect storms, other planes; determine location, altitude; air traffic control, prevent "runway incursions."

FIGURE 9.2 **Example of Oral Presentation Notes**

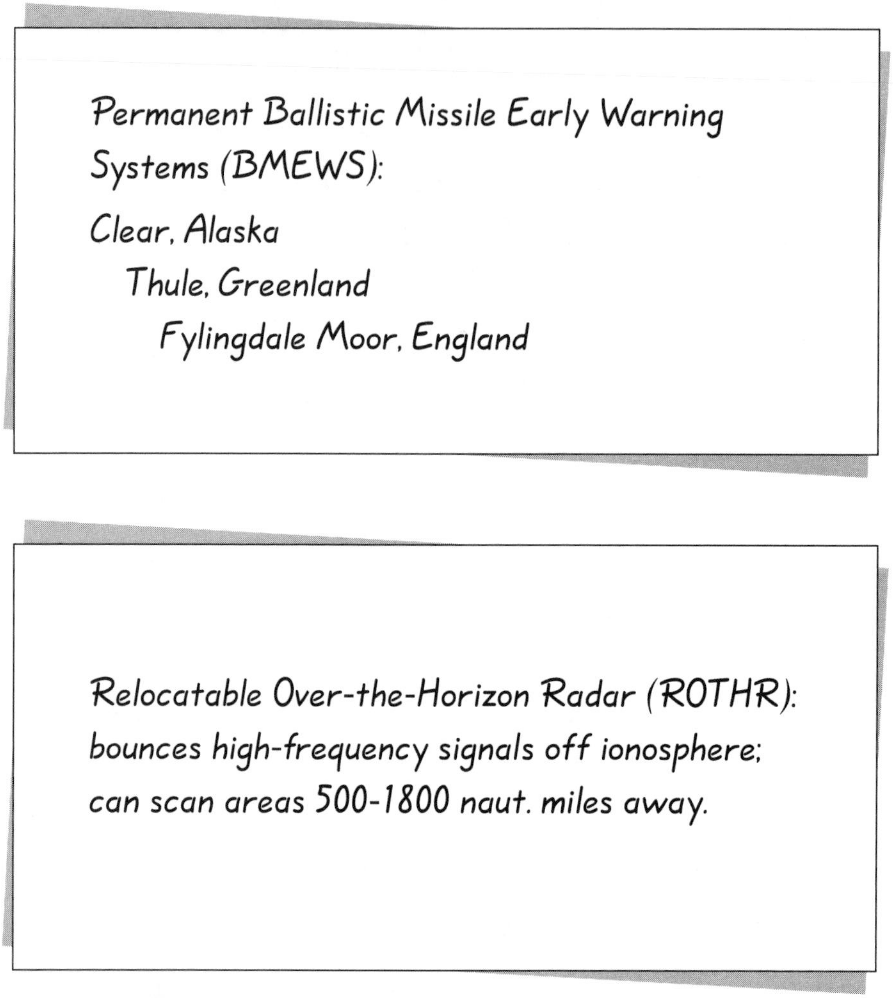

Permanent Ballistic Missile Early Warning Systems (BMEWS):
Clear, Alaska
Thule, Greenland
Fylingdale Moor, England

Relocatable Over-the-Horizon Radar (ROTHR):
bounces high-frequency signals off ionosphere;
can scan areas 500-1800 naut. miles away.

FIGURE 9.3 **Examples of Note Cards**

Rehearsal

Important as the preliminaries are, rehearsal is the most important part of your preparation. Many people skip this step, figuring they'll wing it when the time comes, relying on their wits. Unless you are a very experienced speaker, however, this almost never works. Before attempting to deliver an oral presentation, you *must* practice it. You need not recruit a practice audience (although it certainly helps!), but you must at least recite the speech aloud several times. This will reveal which parts of the presentation seem the most difficult to deliver, and will also establish how *long* the speech really is. You do not want to run noticeably shorter or longer than the allotted time, since this violates the audience's expectations. Remember that speeches tend to run shorter in actuality than in rehearsal, because the pressures of live performance generally speed up the delivery. If aiming for a five-minute presentation, you need seven or eight minutes in rehearsal. If you are expected to speak for half an hour, your rehearsal might take forty to forty-five minutes.

In addition to preparing your speech, you must also prepare *yourself.* All the common sense advice presented in Chapter 8 concerning the employment interview applies equally here. Get a good night's sleep. Shower. Eat, but do not consume any alcoholic beverages. Dismiss any troubling thoughts from your mind. Wear clothing appropriate for the occasion. All of this will contribute to your general sense of confidence and well-being, thereby helping you develop composure and deliver the presentation to the best of your ability.

Delivery

The key to successfully delivering your oral presentation in public is to relax. Admittedly, this is more easily said than done, yet is not as difficult as it may seem. Most audiences are at least reasonably receptive, so you need not fear them. In the classroom setting, for example, all your listeners will soon be called upon to present their own orals, or will have done so already. This usually makes them sympathetic and supportive. It is simply not true that everyone in the room is scrutinizing your every word and gesture, hoping you will perform poorly. At any given moment, in fact, a certain percentage of the audience is probably not paying attention at all. Nevertheless, there are several areas of concern you may wish to consider when delivering an oral presentation.

Introductions and Conclusions

Since first impressions are so important, a good oral presentation must begin with an effective introduction. Here are four useful strategies for opening your speech.

- ***Ask the Audience a Pertinent Question.*** This is an effective introduction because it immediately establishes a connection between you and your listeners—especially if somebody responds. But even if no one does, you can provide the answer yourself, thereby leading smoothly into your discussion. In a presentation on "Tourist

Attractions in New York City," for example, you might open with the query, "Does anyone here know the name of the street the Empire State Building is on?"

- ***Describe a Situation.*** There is something in human nature that makes us love a story, especially if it involves conflict. The enduring appeal of fairy tales, myths and legends, and even soap operas and sentimental country-western lyrics proves the point. You can capitalize on this aspect of your listeners' collective psychology by opening your presentation with a brief story that somehow relates to your subject. A speaker attempting to explore the dangers of tobacco, for example, might begin like this: "My friend Jane, a wonderful young woman with a bright future, had been smoking a pack a day since tenth grade. Finally, at age 25, she had decided to quit. But when she went to the doctor for her annual physical she learned that it was already too late. Tragically, Jane died of lung cancer less than a year later."

- ***Present an Interesting Fact or Statistic.*** This will help you grab the audience's attention by demonstrating that you are familiar with your topic. The annual edition of *World Almanac and Book of Facts* is an excellent source of statistical information on diverse topics, but there are many others as well. Any qualified librarian can direct you to government documents, corporate reports, computer databases, and other useful resources. Even though statistics can be deceptive, people like what they perceive as the hard reality of such data, and therefore find numbers quite persuasive. A speech intended to demonstrate the need for stricter gun control legislation, for example, might open with the observation, "In 1996 alone, there were more than 10,000 handgun-related murders in the United States."

- ***Use a Quotation.*** Get a "Big Name"—Shakespeare, Martin Luther King, the Bible—to speak for you. Find an appropriate saying that will launch your own remarks with flair. Many useful books of quotations exist, but *Bartlett's Familiar Quotations* (available in virtually any good bookstore or library) is the best-known, and for good reason. Bartlett includes nearly 100 quotes on the subject of "money" alone, for example.

The conclusion to your talk is as important as the introduction. Always sum up when you reach the end of an oral presentation. Repeat your key points and show clearly how they support your conclusion. But like an airplane rolling smoothly to a stop on the runway rather than crashing to the ground after reaching its destination, you should not end abruptly. You can accomplish this by returning the audience to the starting point. When you reach the end of your speech, refer to the question, scenario, fact, statistic, or quote with which you opened. This creates in your listeners the satisfying sense of having come full-circle, returning them to familiar territory.

Another common concluding tactic is to ask whether members of your audience have any questions. If so, you can answer them, and then your work is done. If no questions are forthcoming, the audience has in effect ended the speech *for* you. Since this creates the sense of a letdown, however, you can instead have an accomplice or two in the audience ask questions to which you have prepared responses in advance. Although staged, this is a common practice among professional speakers. Whatever form of conclusion you choose, always close by thanking the audience for their time and attention.

Vocal Factors

Obviously, the *voice* is the principal instrument of any oral presentation. Therefore, pay attention to your vocal qualities. Speak at a normal rate of speed, neither too fast nor too slow, and at a normal volume—neither too loud nor too soft. Pronounce clearly, so the audience can understand your every word without undue effort. When using a microphone, be sure it is approximately one foot away from your mouth—any farther, it may not pick up your voice adequately; any closer, your overly amplified b's and p's may create an annoyingly explosive sound. In addition, try to maintain the normal rhythms of everyday conversation. Nothing is more boring than listening to a speech delivered in an unvarying monotone. Conversely, it is irritating to be subjected to an overly theatrical delivery characterized by elaborate gestures or exaggerated vocal effects. The key is to be natural, as if you were speaking to one or two people rather than a whole group.

At the same time, however, an oral presentation is certainly a more formal speaking situation than a social conversation. Therefore, you should provide more examples and illustrations than you ordinarily might, along with more transitional phrases than usual. In addition, make a conscious effort to minimize verbal "ticks," those distracting little mannerisms that characterize everyday speech: "um," "y'know," "okay," "right," and the like. Listening to a tape recording of your oral presentation will enable you to assess the degree to which you need to work on this. While you do not want to sound stiffly artificial, you should also stay away from the more colorful vernacular. Avoid slang, expletives, and conspicuously sub-standard—"I ain't got no"—grammar. Achieving the right level of formality can be challenging, but practicing the presentation a few times will surely help.

Physical Factors

Although your voice is obviously important, your audience *sees* you as well as hears you. They respond to your body language as much as to your words. As you would in an employment interview, you must therefore create a favorable physical impression. Get rid of any chewing gum or tobacco long before stepping up to the lectern. Stand up straight behind the lectern; don't slump or lean over it. Control your hand motions. Do not fold your arms, drum with your fingertips, click a ballpoint pen, or cling rigidly to the lectern with a stiff-armed, white-knuckled grip. Refrain from touching your face or hair, tugging at your clothes, or scratching your body. You can gesture occasionally to make a point, but only if such movements are spontaneous, as in casual conversation. In short, your hands should not distract the audience from what you are saying. Your feet, too, can create problems. Resist the tendency to tap your feet, shift from one leg to the other, or stray purposelessly from the lectern. Plant your feet firmly on the floor and stay put.

In the academic setting, your professors (much like many workplace supervisors) may impose certain regulations concerning proper attire for oral presentations. Baseball caps, for example, are sometimes prohibited, along with various other style and dress affectations such as those mentioned in the Interview section of Chapter 8. Whether in a college classroom or on the job, you should observe any such guidelines, even if you feel they are overly restrictive.

Eye Contact

As much as possible, *look at* your audience. This is probably the hardest part of public speaking. But it is imperative. Unless you maintain eye contact with audience members, you will lose their attention. Keep your head up and your eyes focused forward. If you find it impossible to actually look at your listeners, fake it. Look instead at desk tops, chair legs, or the back wall. But you must create at least the *illusion* of visual contact.

Holding your listeners' attention is one—although certainly not the only—reason why you should absolutely avoid the dreadful error of simply reading to your audience from the text of your speech. Few practices are more boring, more amateurish, or more destructive of audience-speaker rapport. As mentioned in the section on preparation, you should deliver your presentation from notes or note cards rather than from a polished text, to force yourself to adopt a more conversational manner. But keep your papers or cards out of sight, lying flat on the lectern. Do not distract the audience by nervously shuffling them.

Audiovisuals

To greatly enhance your oral presentation, consider using audiovisual aids in conjunction with the variety of visuals (tables, graphs, charts, pictures) discussed in Chapter 3. Audiovisual tools can be very helpful to both you and your audience by illustrating key points throughout your talk. If the room where you are speaking is equipped with a chalkboard, take advantage of it as appropriate. A flip-chart, which is a giant, easel-mounted pad of paper that you write on with felt-tip markers is another useful option. You may also choose to use large display posters prepared in advance, but you must remember to bring along tape or thumbtacks to secure them for viewing.

Whether using a chalkboard, flip-chart, or poster to make your point, however, remember to position yourself *next to* it, not in front; you must not block the audience's view. Remember also to face the audience rather than the display. Be sure your writing is plainly legible from a distance—write in large, bold strokes, using color for emphasis and incorporating the other design principles outlined in Chapter 4. Make sure your drawings and text are easy to see even from the back of the room. There is a rule of thumb for this: the image must be at least one-sixth as large as the distance from which it will be seen. For example, a graph viewed from thirty feet should be five feet wide.

For lettering, use the following chart:

Distance	Size of Lettering
Up to 10 ft.	¾ in.
20 ft.	1 in.
30 ft.	1¼ in.
40 ft.	1½ in.
50 ft.	1¾ in.
60 ft.	2 in.

Although they require more preparation time, you may want to create slides or transparencies that can be projected onto a screen. They lend your presentation a great deal of credibility by making it much more professional and polished. One advantage is that you can control the size of the images on the screen, enlarging them as necessary to create displays that are easily visible even in a relatively big room. Particularly helpful are the computerized presentation software packages that have become increasingly affordable—and versatile—in recent years. These enable the user to create visuals of unprecedented sophistication, incorporating photographs as well as original images, a tremendous range of colors, three-dimensional effects, and other options.

Here are some basic guidelines to bear in mind when using audiovisual media in conjunction with an oral presentation:

- Make certain beforehand that your slides are properly inserted in their tray. They must not be reversed, upside-down, or out of sequence.
- Do not include too much information on a transparency; keep it simple. If a given transparency is unavoidably information-rich, reveal only a section at a time, sliding a piece of paper downward to uncover each section when you are ready to discuss it. Thus the information on the screen will gradually expand in pace with your commentary, and your audience will be prevented from reading ahead and possibly neglecting your explanation. If you are using presentation software, this principle can be applied electronically.
- To draw the audience's attention to a detail on the screen, use a laser pointer rather than a yardstick or conventional pointer, which are effective only for pointing out details on chalkboards, posters, maps, and the like. If you do not have a laser pointer, highlight details by pointing to them on the transparency itself, but be sure to use a pen or freshly sharpened pencil rather than your finger, because the bulky shadow cast by your hand will block too much from view. Presentation software, of course, affords a variety of far more imaginative highlighting methods.
- When adding notations on a transparency during your presentation, be sure to write legibly, using a marker designed for that purpose. Again, software greatly simplifies this procedure.

- Avoid the glaring "Empty White Screen" effect; turn off your slide or transparency projector once you are done with it, or if you will not be referring to it for more than a minute or two.

Depending on the length, scope, and topic of your speech, you may decide to supplement your remarks with videotape or sound recordings, provided they are of good quality. Relevant physical objects can also be displayed or passed around. If you were explaining how to tune a guitar, for example, you would certainly want to demonstrate the procedure on an actual instrument. Similarly, if you were explaining the workings of a particular tool or other device, ideally you would provide one (or more) for the audience to examine. However, these materials should *supplement*—not overshadow—your remarks. Much like visuals in written reports, audiovisual devices in oral presentations must always serve a useful purpose, enhancing without dominating, and should never be introduced simply for their own sake.

Enthusiasm

Try to deliver your oral presentations in a lively, upbeat, enthusiastic manner. This actually makes your job easier, since a positive attitude on your part will help to foster a more receptive attitude on the part of the audience. If your listeners sense that you would rather be elsewhere, they tend to "tune out." When that occurs, you receive no encouraging feedback, and knowing you've lost your audience makes it even more difficult to continue. If you sense, however, that the audience is following along, this reinforcement in turn fuels your performance. But this cannot happen unless you project in an engaging way. From the start, *you* establish the tone. Therefore, it makes sense to adopt a positive attitude when giving an oral presentation, not only for the audience's sake but to serve your own purposes as well.

The many factors we've examined that contribute to a good delivery may seem like a lot to keep track of. If you are like most speakers, however, you probably have real difficulty in only one or two areas. An especially useful strategy is to videotape your rehearsal to determine what you should work on to improve your delivery. As stated at the outset, a successful oral presentation is the result of preparation, composure, and common sense. If you take seriously the recommendations offered in this chapter, and practice the strategies and techniques suggested, your performance as a public speaker will improve greatly.

✓ Checklist Evaluating a Public Speaker

A good public speaker

___ opens with an interesting, attention-getting introduction;

___ follows a clear and logical pattern of organization;

___ provides enough detail to fully develop the subject;

___ closes with a smooth, satisfying conclusion;

___ speaks in a firm, clear, expressive voice;

___ makes frequent eye contact with the audience;

___ appears physically relaxed and composed, and uses no distracting mannerisms;

___ maintains an appropriate level of formality, neither too casual nor too solemn;

___ delivers in an alert, engaging manner;

___ satisfies but does not exceed the appropriate length for the presentation.

Exercises

■ EXERCISE 9.1

Prepare and deliver a five- to ten-minute oral presentation on one of the following autobiographical topics:

- A Childhood Memory
- My Brush with Danger
- My Angriest Moment
- My Most Satisfying Accomplishment
- My Career Goals
- What I Expect My Life to Be Like in Ten Years

■ EXERCISE 9.2

Prepare and deliver a five- to ten-minute oral presentation that summarizes a book, article, lecture, film, or television broadcast related to your field of study or employment (see Chapter 5).

EXERCISE 9.3

Prepare and deliver a five- to ten-minute oral presentation that provides a specific mechanism description related to your field of study or employment. Present an actual example of such a mechanism, along with any audiovisual aids that may be helpful to your audience (see Chapter 6).

EXERCISE 9.4

Prepare and deliver a five- to ten-minute oral presentation describing a process related to your field of study or employment. Present any audiovisual aids that may be helpful to your audience (see Chapter 6).

EXERCISE 9.5

Prepare and deliver a five- to ten-minute oral presentation describing a procedure related to your field of study or employment. Present any audiovisual aids that may be helpful to your audience (see Chapter 6).

EXERCISE 9.6

Prepare and deliver a five- to ten-minute oral presentation providing instructions related to your field of study or employment. Present any audiovisual aids that may be helpful to your audience (see Chapter 7).

EXERCISE 9.7

Prepare and deliver a five- to ten-minute oral presentation based on Exercise 4.10. Present any audiovisual aids that may be helpful to your audience.

EXERCISE 9.8

Prepare and deliver a five- to ten-minute oral presentation based on Exercise 6.18. Present any audiovisual aids that may be helpful to your audience.

EXERCISE 9.9

Prepare and deliver a five- to ten-minute oral presentation based on Exercise 7.8. Present any audiovisual aids that may be helpful to your audience.

EXERCISE 9.10

Prepare and deliver a five- to ten-minute oral presentation based on Exercise 8.4. Present any audiovisual aids that may be helpful to your audience.

10

Long Reports: Format, Collaboration, and Documentation

Learning Objective Upon completing this chapter, you will be able to create well-designed long reports whether working independently or with others, and to correctly document the sources of your information.

In business, industry, and the professions, important decisions are made every day. Some concern routine matters while others are more complicated, involving considerable risk and expense. Let us suppose, for example, that a hospital administration is debating whether to add a new wing to the main building. Or perhaps a police department wants to switch to a different kind of patrol car, or a successful but relatively new business venture must decide whether to expand now or wait a few years. In each case, the situation requires in-depth study before a responsible decision can be reached. The potential advantages and drawbacks of each alternative have to be identified and examined, as well as the long-range effects. This is where the long report comes into play. This chapter discusses how to prepare such a report, explaining its formatting components, the dynamics of group-written reports, and some standard procedures for documenting sources.

Format

Obviously, both the subject matter and the formatting of long reports will vary from one workplace to another and in the academic context, from one discipline to another, and even among instructors. Nevertheless, most long reports share certain components. Let us consider each of these.

Transmittal Document

Prepared according to the standard memo or business letter format (see Chapter 2), the transmittal document accompanies a long report, conveying it from whoever wrote it to whoever requested it. The transmittal document says, in effect, "Here's that report you wanted," and very briefly summarizes its content. The memo format is used for transmitting in-house reports, whereas the letter format is used for transmitting reports to outside readers. Often the transmittal document serves as a "cover sheet," although sometimes it is positioned immediately after the title page of the report. See Figure 10.1, page 240, for a sample transmittal memo.

Title Page

In addition to the title itself, this page includes the name(s) of whoever prepared the report, the name(s) of whoever requested it, the names of the companies or organizations involved, and the date. In an academic context, the title page includes the title, the names of the student author(s) and the instructor who assigned the report, the course name (along with the course number and section number), the college, and the date. See Figure 10.1, page 241, for a sample title page prepared for a workplace context.

Abstract

Sometimes called an "executive summary," this is simply a brief synopsis—a greatly abbreviated version of the report (see Chapter 5). An effective abstract captures the essence of the report, including its major findings and recommendations. In the workplace, the function of an abstract is to assist those who may not have time to read the entire report, yet need to know what it says. Sometimes the abstract is positioned near the front of the report, other times it appears at the end. For a ten- or twenty-page report, the abstract should not be longer than one page, and can be formatted as one long paragraph. Figure 10.1, page 242, offers an example of a concise abstract.

Table of Contents

As in a book, the table of contents for a long report clearly shows each numbered section of the report, along with its title and the page on which it appears. Many also show subdivisions within sections. When fine-tuning a report before submitting it, check to ensure that the section numbers, titles, and page numbers used in the table of contents are consistent with those in the report itself (see Figure 10.1, page 243).

List of Illustrations

This list resembles the table of contents, but rather than referring to text sections, it lists tables, graphs, charts, and all other visuals appearing in the report—each numbered and titled—and their page numbers. As with the table of contents, always check to ensure that your illustrations list accurately reflects the visual contents of the report and their labeling/captions (see Figure 10.1, page 244).

Glossary

A "mini-dictionary," the glossary defines all potentially unfamiliar words, expressions, or symbols in your report. Not all reports need a glossary; it depends on the topic and the intended audience. But if you are using specialized vocabulary or symbols that may not be well known, it is best to include a glossary page with terms alphabetized for easy reference, and symbols listed in the order in which they appear in the text (see Figure 10.1, page 245).

Text

One major difference between a long report and an academic term paper is that a report is divided into sections, usually numbered, and each with its own title. As mentioned previously, it is important that these divisions within the text be accurately reflected in the table of contents.

Every long report also includes an introduction and a conclusion. The introduction provides an overview of the report, identifying its purpose and scope, and explaining the procedures used and the context in which it was written (see Figure 10.1, page 246). The conclusion summarizes the main points in the report and lists recommendations, if any (see Figure 10.1, page 253).

Visuals

A major feature of many reports (see Chapter 3), visuals sometimes appear in a separate section—an "appendix"—at the end of a report. A better approach, however, is to integrate them into the text, as this is more convenient for the reader. Either way, you should draw the reader's attention to pertinent visuals (stating, for example, "See Figure 5"), and every visual must be properly numbered and titled, with its source identified. The numbering/titling system must be the same system used in the list of illustrations.

Pagination

Number your report pages correctly. There are several pagination systems in use. Generally, page numbers (1, 2, 3, and so on) begin on the introduction page and continue on until the last page of the report. "Front matter" pages (abstract, table of contents, list of illustrations, glossary, and anything else that precedes the introduction) are numbered with lowercase Roman numerals (i, ii, iii, iv, v, vi, and so on). There is no page number on the transmittal document or the title page, although the latter "counts" as a front-matter page; hence the page immediately following the title page is numbered as "ii." The best position for page numbers is in the upper right-hand corner, because that location enables the reader to find a particular page simply by thumbing through the report. Notice the page numbering throughout Figure 10.1.

Collaboration

A memo, letter, or short report nearly always is composed by one person working individually. This is sometimes true of long reports as well. However, since the subject matter of long reports is often complex and multifaceted, they are often written collaboratively. This kind of teamwork is very common in the workplace because it provides certain obvious advantages. For example, a group that works well together can produce a long report *faster* than one person working alone. In addition, the team possesses a broader perspective and a greater range of knowledge and expertise than an individual. To slightly amend the old saying, two heads—or more—are obviously better than one.

Nevertheless, collaboration can sometimes be problematic because the members of a given group may find it difficult to interact smoothly. Real teamwork requires everyone involved to exercise tact, courtesy, and responsibility. The following factors are essential to successful, rewarding collaboration.

1. Everyone on the team must fully understand the purpose and goals of the project, and agree to set aside individual preferences in favor of the group's collective judgment.

2. A team leader is placed in charge of the project—someone whom the other members are willing to recognize as the coordinator. Therefore, this leader should be *elected* from within the group. The leader must, of course, be knowledgeable and competent but must also be a "people person" with excellent interpersonal skills. Once the group has clearly identified the goals and scope of the project, the leader firmly enforces the agreed-upon rules, negotiates and resolves conflict, and holds the whole effort together.

3. The team next assigns clearly defined roles to the other members, designating responsibilities according to everyone's talents and strengths. For example, the group's most competent researcher takes charge of information retrieval. Someone trained in drafting or computer-assisted design formats the report and creates visuals. The member with the best keyboarding skills (or clerical support) produces the actual document. The best writer is the overall editor,

making final judgments on matters of organization, style, mechanics, and the like. If an oral presentation is required, the group's most confident public speaker assumes that responsibility. A given individual may play more than one role, but everyone must feel satisfied that the work has been fairly distributed.

4. Once the project is under way, the team meets regularly to assess each other's progress, prevent duplication of effort, and resolve any problems that may arise. All disagreements or differences of opinion are reconciled in a productive manner. In any group undertaking, a certain amount of conflict is inevitable and indeed necessary to achieve consensus. But this interplay should be a source of creative energy, not antagonism. Issues must be dealt with on an objectively intellectual level, not on a personal or emotional level. To this end, the group should adopt a code of interaction designed to minimize conflict and maximize the benefits of collaboration. Each team member should:

   - make a real effort to be calm, patient, reasonable, flexible—in short, *helpful;*
   - voice all reservations, misgivings, and resentments rather than let them smolder;
   - direct criticism at the *issue,* not the person—"There's another way of looking at this" rather than "You're only looking at this one way"—and try not to interpret criticism personally;
   - avoid interrupting when others are speaking;
   - paraphrase others' remarks—"What you're saying, then, is . . ."—to make certain of their meaning;
   - suggest rather than command—"Maybe we should try it this way" instead of "Do it this way!"—and offer rather than demand—"If you'd like, I'll . . . ," instead of "I'm going to . . .";
   - accentuate the positive rather than the negative—"Now that we've agreed on the visuals, let's move on" instead of "We can't seem to agree on anything but the visuals."

5. All members of the group must complete their fair share of the work in a conscientious fashion and observe all deadlines. Nothing is more disruptive to a team's progress than an irresponsible member who fails to complete work punctually, or "vanishes" for long periods of time. To contact one another between regularly scheduled meetings of the group, members should exchange telephone numbers and/or e-mail addresses. Indeed, the team leader should monitor the members' progress, reminding them of upcoming deadlines and meetings.

The group can handle the actual writing of the report in any of several ways:

- The whole team writes the report collectively, then the editor revises the draft and submits it to the group for final approval or additional revisions.
- One person writes the entire report, then the group—led by the editor—revises it collectively.
- Each team member writes one part of the report individually, then the editor revises each part and submits the complete draft to the whole group for final approval or additional revisions.

Of these alternatives, the first is the most truly collaborative, but is also extremely difficult and time-consuming, requiring uncommon harmony within the group. The second method is preferable, but places too great a burden on one writer. The third approach, therefore, is usually the best, provided the editor seeks clarification from individuals whenever necessary during the editing process. Note, however, that in every case the whole group gets to see and comment on the report in its final form. Since everyone's name will be on it, there should be no surprises when the finished product is released. "Collaboration" should be precisely that, producing a polished report approved by all members of the team.

Documentation

Documentation is simply a technical term for the procedure whereby writers identify the sources of their information. In the workplace and in popular periodicals, this is often accomplished by inserting the pertinent information directly into the text, as in this example:

> As Ingersoll-Rand quality assurance manager Robert A. Wenderlich explains in his article "Intake Details Affect Compressor System Performance" in the January 1994 issue of *Maintenance Technology*, a compressor inlet filter "should be located away from areas of high humidity and temperature such as drying areas, evaporative cooling towers, or other such high-humidity or high-vapor areas" (p. 23).

This straightforward approach eliminates the need for a bibliography (list of sources) at the end of the piece. Documentation in academic writing, however, nearly always includes both a bibliography and parenthetical citations identifying the origin of each quotation, statistic, paraphrase, or visual within the text.

Bibliography

There are several standard ways to format a list of citations. The Modern Language Association (MLA) format, which titles the list "Works Cited," and the American Psychological Association (APA) format, which titles the list "References," are among the most widely used in academic settings. A typical bibliography entry under each system would look like this:

MLA Sacks, Oliver. *Seeing Voices: A Journey into the World of the Deaf.* Berkeley: U of California P, 1989.

APA Sacks, O. (1989). *Seeing voices: A journey into the world of the deaf.* Berkeley: University of California Press.

Notice that in MLA format the second and any subsequent lines of each entry are indented, while in APA format only the *first* line is indented. Perhaps the most obvious difference between the two formats is the placement of the date of publication. Differences also exist with respect to capitalization, punctuation, and abbreviation. In both formats, however, double-spacing is used throughout, and book titles—like the titles of newspapers, magazines, journals, and other periodicals—are italicized. In both formats, entries appear in alphabetical order by author's last name or, in the case of an anonymous work, by the first word of the title.

There are many other kinds of sources besides a single-author book, however, and each requires a slightly different handling. Some of the most common citations follow.

Book by Two Authors

MLA Rupp, Leila J., and Verta Taylor. *Survival in the Doldrums: The American Women's Rights Movement, 1945 to the 1960s.* New York: Oxford UP, 1987.

APA Rupp, L. J., & Taylor, V. (1987). *Survival in the doldrums: The American women's rights movement, 1945 to the 1960s.* New York: Oxford University Press.

Book by Three Authors

MLA Gradwell, John, Malcolm Welch, and Eugene Martin. *Technology: Shaping Our World.* South Holland, IL: Goodheart, 1993.

APA Gradwell, J., & Welch, M., & Martin, M. (1993). *Technology: shaping our world.* South Holland, IL: Goodheart-Willcox.

Book by More Than Three Authors

MLA Althouse, Andrew D., et al. *Modern Welding.* South Holland, IL: Goodheart, 1992.

APA Althouse, A. D., Turnquist, C. H., Bowditch, W. A., & Bowditch, K. E. (1992). *Modern welding.* South Holland, IL: Goodheart-Willcox.

Edited Book

MLA Skolnik, Merrill I., ed. *Radar Handbook.* New York: McGraw, 1970.

APA Skolnik, M. I. (Ed.). (1970). *Radar handbook.* New York: McGraw-Hill.

Book by a Corporate Author

MLA Air Conditioning and Refrigeration Institute. *Refrigeration and Air Conditioning.* Englewood Cliffs, NJ: Prentice, 1979.

APA Air Conditioning and Refrigeration Institute (1979). *Refrigeration and air conditioning.* Englewood Cliffs, NJ: Prentice-Hall.

Article in a Newspaper

MLA Leary, Warren. "AIDS and Tuberculosis Epidemics Rise in Deadly Combination." *New York Times* 4 June 1995: 8.

APA Leary, W. (1995, June 4). AIDS and tuberculosis epidemics rise in deadly combination. *New York Times,* p. 8.

To document a print source that you have accessed through a reputable on-line database, you must provide additional information at the end of the entry: the name of the database, the name of the index, and the date of access. Each of the above entries would therefore conclude like this:

New York Times Online. Online. Nexis. 1 July 1995.

Article in a Weekly or Biweekly Magazine

MLA Witkin, Gordon. "A New Assault on Cocaine." *U.S. News & World Report* 11 Jan. 1993: 21.

APA Witkin, G. (1993, Jan. 11). A new assault on cocaine. *U.S. News & World Report,* p. 21.

Article in a Monthly or Bimonthly Magazine

MLA Simanaitis, Dennis. "Synthetic Oils." *Road & Track* Feb. 1993: 98–101.

APA Simanaitis, D. (1993, Feb.). Synthetic oils. *Road & Track,* pp. 98–101.

Article in a Trade Journal or Academic Journal

MLA Sondey, Margaret. "An Initial Investigation of Welded Homes in the United States." *Welding Innovation Quarterly* 9. 2 (1992): 4–7.

APA Sondey, M. (1992). An initial investigation of welded homes in the United States. *Welding Innovation Quarterly,* 9 (2), pp. 4–7.

In both formats the numbers immediately following the journal's title identify the volume (9) and the issue (2). If the journal is paginated continuously by volume, the issue number is omitted from the bibliography entry.

Article in an Edited Collection of Articles

MLA Chatterton, Michael R. "Police Work and Assault Charges." *Control in the Police Organization.* Ed. Maurice Punch. Cambridge: MIT P, 1983. 194–221.

APA Chatterton, M. R. (1983). Police work and assault charges. In M. Punch (Ed.), *Control in the police organization* (pp. 194–221). Cambridge: Massachusetts Institute of Technology Press.

Anonymous Article

MLA "Hairy Carbon Extends Battery Life." *Popular Science* Jan. 1993: 25.

APA Hairy carbon extends battery life. (1993, January). *Popular Science,* p. 25.

Entry in Encyclopedia or Other Reference Work

MLA Visich, Marian. "Turbine." *World Book Encyclopedia,* 1986 ed.

APA Visich, M. Turbine. (1986). *World Book Encyclopedia.*

Material from a Portable Electronic Database

MLA Engelhardt, A. G., and M. Kristiansen. "Ohm's Law." *New Grolier Multimedia Encyclopedia for Macintosh.* CD-ROM. Grolier, 1993.

APA Engelhardt, A. G., & Kristiansen, M. (1993). Ohm's law. *New Grolier multimedia encyclopedia for Macintosh.* CD-ROM. Grolier.

Material from an On-Line Electronic Source

MLA "United Nations Conference." United Nations Conference on Trade and Development. On-line. Internet. 14 Aug. 1996. Available WWW: http://gatekeeper.unicc.org/unctad/.

APA *United Nations conference* [On-line WWW]. (n.d.) United Nations Conference on Trade and Development. Available http://gatekeeper.unicc.org/unctad/.

In terms of credibility, online sources span a very broad range. Much of the information available on the Internet derives from reliable authorities. A great deal more, however, does not. The fact that information has been accessed via a well-known service provider such as AOL, CompuServ, or Netcom guarantees absolutely nothing about the validity of that material. Since the Internet is a completely democratic medium, anyone can post anything at all, no matter how unverifiable. For this reason, you must exercise caution when using electronic sources in academic research. Indeed, many instructors issue strict guidelines regarding this matter, and most will not accept information received from casual forums such as chat rooms, bulletin boards, and the like. When in doubt about the acceptability of an on-line source, always check with your instructor. Since documentation procedures for electronic sources are—like the technology itself—in a state of continuous evolution, you would be well advised to consult the MLA's own Web site <http://www.mla.org> for up-to-date guidelines.

Radio or Television Broadcast

MLA "The Broken Cord." Interview with Louise Erdrich and Michael Dorris. Dir. and prod. Catherine Tatge. *A World of Ideas with Bill Moyers.* Exec. prod. Judith Davidson Moyers and Bill Moyers. Public Affairs TV. WNET, New York. 27 May 1990.

APA Erdrich, L., & Dorris, M. (1990, May 27). The broken cord [Interview]. *A World of ideas with Bill Moyers* [Television program]. New York: Public Affairs TV. WNET.

Personal Interview

MLA Smith, Ivan. Personal interview. 10 Nov. 1999.

APA APA style requires that all such sources (personal or telephone conversations, interviews, and the like) be documented within your text, as follows:

Economist Ivan Smith (personal conversation, November 10, 1999) stated that the "total cost of the project will be well over a million dollars."

Parenthetical Citations

Every time you use one of your sources within the body of a report, whether quoting directly or paraphrasing in your own words, you must identify the source by inserting parentheses. The contents and positioning of the parentheses vary somewhat depending on whether you're using the MLA or APA format. Here is an example of each variation:

MLA "To be deaf, to be born deaf, places one in an extraordinary situation" (Sacks 116).

APA "To be deaf, to be born deaf, places one in an extraordinary situation" (Sacks, 1989, p. 116).

If you mention the author's name in your own text, neither MLA nor APA requires that the name appear in the parentheses, although the APA system then requires *two* parenthetical notations:

MLA As Sacks observes, "to be deaf, to be born deaf, places one in an extraordinary situation" (116).

APA As Sacks (1989) observes, "to be deaf, to be born deaf, places one in an extraordinary situation" (p. 116).

When you are paraphrasing, the differences between the two systems are the same as when you are quoting.

MLA Deafness puts a person in unusual circumstances (Sacks 116).

APA Deafness puts a person in unusual circumstances (Sacks, 1989, p. 116).

MLA As Sacks points out, deafness puts a person in unusual circumstances (116).

APA As Sacks (1989) points out, deafness puts a person in unusual circumstances (p. 116).

To credit a quote or a paraphrase from an unsigned source (such as the "Anonymous Article" example shown on page 236), parenthesize the title (or a shortened version of it), along with the page number.

MLA Engineers have developed a new material called "hairy carbon," which has superior electrical conductivity ("Hairy" 25).

APA Engineers have developed a new material called "hairy carbon," which has superior electrical conductivity (Hairy, p. 25).

The purpose of parenthetical citations is to enable readers to find your sources on the "Works Cited" or "References" page, in case they wish to consult the original sources.

Obviously, proper documentation is an important part of any report or other paper that has drawn on sources beyond the writer's own prior knowledge. Figure 10.1 presents a correctly prepared report, "Drug Testing in the Workplace," with sources documented according to MLA guidelines. As mentioned earlier, actual reports written in the workplace may employ other kinds of documentation. But certainly the *format* of this report is fairly typical of the kind used in the workplace, and in most college English courses focusing on workplace communications. Once you have mastered format, you can adapt it to a wide range of situations, whatever documentation system you use.

PARAMOUNT MANUFACTURING, INC.

MEMORANDUM

DATE: December 1, 1999

TO: Rosa Sheridan
Director of Human Resources

FROM: William Congreve
Administrative Assistant

SUBJECT: Drug Testing Report

As you may recall, we decided last month that I should compile a report on the current status of drug testing programs in the American workplace, so that we might explore the feasibility of introducing such a program here at Paramount Manufacturing. Here is the report. If you have any questions, I shall be happy to provide further details.

FIGURE 10.1 **Drug Testing Report, Transmittal Memo**

DRUG TESTING
IN THE WORKPLACE

by

William Congreve
Administrative Assistant

Submitted to

Rosa Sheridan
Director of Human Resources

Paragon Manufacturing, Inc.
Mission Viejo, California

December 1, 1999

FIGURE 10.1 **Drug Testing Report, Title Page**

ii

ABSTRACT

Paragon Manufacturing, Inc. is considering introducing a mandatory drug testing program. Although intended to reduce the costs associated with workplace substance abuse, drug testing is quite controversial. Some experts argue that the extent of workplace drug abuse has been greatly exaggerated, and that drug programs—first introduced in large numbers in the 1980s—are a needless violation of employees' privacy. Most drug testing programs rely on EMIT, a test that often yields inaccurate results, thus necessitating the use of confirmatory GC/MS tests, to reduce the possibility of false positives (and false negatives). Drug testing appears to be least problematic when used to screen applicants for employment rather than established employees. To avoid costly lawsuits and other setbacks, progressive companies observe several key features of successful drug testing protocol: a clearcut policy statement, strict guidelines for specimen collection, use of NIDA-certified laboratories, confirmation of all positive test results, and employee assistance services. For Paragon to introduce a testing program, the best path may be to begin by testing job applicants only, rather than the existing workforce. In any case, however, more study is needed, preferably with the assistance of an outside (NIDA) consultant.

FIGURE 10.1 **Drug Testing Report, Abstract**

iii

TABLE OF CONTENTS

FIGURE 10.1 Drug Testing Report, Table of Contents

iv

LIST OF ILLUSTRATIONS

FIGURE 10.1 **Drug Testing Report, List of Illustrations**

v

GLOSSARY

antibodies: Protein substances developed by the body, usually in response to the presence of antigens (bacteria, for example).

assay: Analysis of a substance to determine its constituents and the relative proportions of each.

enzymes: Organic catalyst produced by living cells but capable of acting independently; complex colloidal substances, they can induce chemical changes in other substances without undergoing change themselves.

false positive: Test result that incorrectly indicates the presence of the substance(s) tested for.

false negative: Test result that incorrectly indicates the absence of the substance(s) tested for.

mass spectrum: Identifiable pattern of electromagnetic energy given off by a substance under specific test conditions.

metabolites: Products of metabolism.

silica: Silicon dioxide, SiO_2.

FIGURE 10.1 Drug Testing Report, Glossary

I - INTRODUCTION

Since the founding of the company in 1952, Paragon Manufacturing, Inc. has always sought to achieve maximum productivity while providing a safe, secure, and conducive work environment for our employees. In keeping with those goals, management has determined that it may now be time for Paragon to take a more active role in the war against drugs, by adopting measures to ensure a drug-free workplace. One such measure that has been suggested is the creation of a mandatory drug testing policy for all new and established employees, and this idea is currently under consideration.

Since drug testing is somewhat controversial, however, there is need for further study and careful deliberation before a determination is reached. This report, compiled after an in-depth review of recent professional literature on the subject, is intended as a first step in that process.

II - BACKGROUND

Various estimates of the annual costs of workplace drug abuse differ greatly, but all the numbers are in the billions, with the highest exceeding $160 billion per year in "accident related loss of life and property and reduced productivity" (West and Ackerman 584). In 1994 the National Safety Council put the figure at $111 billion, broken down into six main categories, as shown in Table 1, below.

Table 1 Work-Injury Costs

Wage and Productivity Losses	$59.9 billion
Medical Costs	20.7 billion
Administrative Expenses	14.4 billion
Employer Costs	9.7 billion
Damage to Motor Vehicles	4.1 billion
Fire Losses	3.1 billion
Total	$111.9 billion

Source: "An Alternative to Drug Testing?"

FIGURE 10.1 **Drug Testing Report, page 1**

2

But several recent studies contend that the majority of these workplace injuries are attributable to causes other than substance abuse. As shown in Table 2, next page, "Dangerous working conditions, noise and dirt on-the-job, and conflicts at work appear to be the greatest predictors of job injuries. Sleeping problems, which may be exacerbated by shift work, also seem likely to be another direct cause of job injuries" (Macdonald 718). In addition, other recent reports maintain that substance abuse in society at large—and, by implication, in the workplace—is actually decreasing even while the issue becomes ever more exaggerated. "At a national and local level, politicians, government agencies, law enforcement agencies, and the media directly benefit from concern over drugs and . . . find it in their best interests to sensationalize drug abuse" (Crow and Hartman 928).

In any case, workplace drug testing programs are quite common. According to the Bureau of Labor Statistics, 67% of large companies (5,000 employees or more) maintain programs (Brookler 128), while many medium-size companies also do. A large number of these programs were introduced in the 1980s, a period of sharply increasing drug use (Konovsky and Cropanzano 698). One response to that increase was the report from the President's Commission on Organized Crime, *America's Habit: Drug Abuse, Drug Trafficking and Organized Crime* (1986), which recommended that "Government and private sector employers who do not already require drug testing of job applicants and current employees should consider the appropriateness of such a testing program" (Glantz 1427).

In that same year, President Reagan signed Executive Order 12564, which mandated drug testing for federal employees in "safety sensitive" positions (law enforcement, public health and safety, and national security, for example) and for federally regulated industries such as transportation and nuclear power (Payson and Rosen 25). State and local governments soon followed suit, especially in the police and corrections sectors, as did private employers. This trend continued under President Bush's National Drug Control Strategy and his White House mandate for "fair but tough drug-free workplace programs in the private sector" (Brookler 132). Not everyone has been supportive, however, and some drug testing cases have been challenged on Constitutional grounds, specifically the Fourth Amendment protection

FIGURE 10.1 Drug Testing Report, page 2

Table 2 Percentage of Total Job Injuries Associated With Each Variable

Variable		Injuries % (N)	Total N	Percentage of Injuries Associated With Variable
Trouble Sleeping	NO	5.7 (12)	210	77.8
	YES	15.4 (42)	273	
Noise and Dirt	NO	6.2 (20)	324	62.3
	YES	22.0 (33)	150	
Danger	NO	6.8 (26)	380	50.9
	YES	28.1 (27)	96	
Shift Work	NO	8.2 (31)	365	41.5
	YES	21.8 (22)	101	
Worry	NO	9.0 (35)	387	34.0
	YES	19.8 (18)	91	
Boredom	NO	9.1 (37)	408	28.8
	YES	21.7 (15)	69	
Conflict	NO	9.1 (42)	419	28.3
	YES	25.0 (15)	60	
Illicit Drug Use	NO	9.8 (43)	440	20.4
	YES	25.6 (11)	43	

Source: Macdonald, p. 711

against unreasonable search and seizure (Hukill 75). Indeed, the city of San Francisco (among others) has an ordinance prohibiting private-sector employers from administering drug tests except under specific conditions, with no random or company-wide screening permitted (Payson and Rosen 30).

III - TECHNICAL ASPECTS

Widespread drug testing first became possible in the 1970s, with the development of a relatively low-cost chemical assay called EMIT (enzyme multiplied immunoassay test), which can be used to detect the presence of a

FIGURE 10.1 Drug Testing Report, page 3

broad spectrum of drugs (alcohol, steroids, stimulants, sedatives, opiates, hallucinogens, and others) in a single specimen of urine (West and Ackerman 579). Originally used only by crime laboratories, treatment centers, and the military, this technology quickly spread to the business and manufacturing sector (O'Keefe 34), where it is used to screen both prospective and current employees for evidence of substance abuse.

As Zimmer and Jacobs explain, urinalysis detects not the actual drug consumed, but the metabolites created by its processing in the system. The EMIT technique introduces into the urine specimen a sample of the metabolite being sought, with a detectable enzyme "tag" attached to each of its molecules. Antibodies that bind to the metabolites are also added. If the specimen was originally free of drug metabolites, the antibodies will bond to the tagged ones, creating a new substance in which the "tag" is no longer detectable. But if the specimen *does* contain drug metabolites, some of those molecules will bind to the antibodies, leaving a surplus of tagged molecules that are unattached and therefore detectable (Zimmer and Jacobs 4–5).

A valuable feature of this technique is its adaptability to automation. Using sophisticated equipment such as the Olympus AU-500, a laboratory technician is able to analyze 4,000 urine samples in an hour. Hitachi's model 736-50 can almost double that pace (Zimmer and Jacobs 5). Hence employers can expect relatively fast turnaround, always an important consideration in personnel matters.

For a variety of reasons, however, test results are often inaccurate. One study conducted by the Centers for Disease Control (CDC) found accuracy rates as low as 33 percent, while a Northwestern University study reported 25 percent of the positive findings resulting from the EMIT technique were in fact "false positives." Indeed, EMIT's manufacturer recommends that all positive findings be confirmed by means of more sophisticated follow-up tests (Chineson 91).

Among the most reliable of the confirmatory tests is a method known as gas chromatography/mass spectrometry (GC/MS). In this procedure, the urine sample is vaporized and then forced through a silica-coated tube. Because silica slows down substances at different rates, they exit the tube individually.

FIGURE 10.1 Drug Testing Report, page 4

Ion bombardment of the exiting substances causes them to fragment in patterns that can then be matched to the known mass spectrum of each drug. Although expensive, the GC/MS procedure is very accurate. "Routine GC/MS confirmation eliminates most false positives and allows laboratories to set immunoassay cutoff levels lower, thereby also reducing the number of false-negative results" (Zimmer and Jacobs 6–7).

IV - LEGAL AND ETHICAL ASPECTS

Despite drug testing's obvious appeal to employers wishing to rid the workplace of controlled substances, a number of legal and ethical problems surround its use. The practice is sometimes seen as "demeaning and intrusive" (Glantz 1427) and, as mentioned earlier, has been contested on Fourth Amendment grounds. The courts have been inconsistent, sometimes ruling in favor of the plaintiff, sometimes not. "The question is always, how far should the balance tip away from the right to be free from . . . intrusion—how weighty must the public interest be to abridge their rights?" (Glantz 1428).

Clearly, this is a difficult question that can be decided only on a case-by-case basis. There is no denying, however, that mandatory drug testing typically requires "individuals to submit to a highly intrusive test to vindicate . . . innocence. Such a presumption of guilt is contrary to the Constitution" (Glantz 1428). In addition, the tests constitute a serious invasion of workers' medical privacy, for urinalysis can reveal not only substance abuse, but also pregnancy, asthma, and the fact that an individual is being treated for heart disease, manic depression, epilepsy, diabetes, and "a host of other physical and mental conditions" (O'Keefe 38).

On the other hand, mandatory drug testing—especially of job applicants—can be an effective means of weeding out those whose use of controlled substances will almost certainly have a negative impact on job performance. Several studies have substantiated this. A Postal Service experiment screened 5,465 job applicants (of whom 4,375 were hired) at 21 participating sites, testing for a variety of drugs. The results were not immediately released, and therefore had no impact on hiring decisions. But

FIGURE 10.1 **Drug Testing Report, page 5**

6

follow-ups revealed that those who had tested positive were about 1.5 times more likely to be fired, and were more than 1.75 times more prone to absenteeism. In the case of cocaine specifically, users were twice as likely to be fired and more than three times as likely to be absent (West and Ackerman 586–587).

In view of such findings, it is hardly surprising that many major companies such as Coors, Georgia Power, and 3M conduct aggressive anti-drug campaigns targeting not only applicants but established employees ("Drug abuse at work"). Similarly, "despite the opposition of the AMA . . . physicians at Johns Hopkins Hospital in Baltimore and at Columbia-Presbyterian Medical Center in New York are . . . required to undergo random . . . screening" (West and Ackerman 589).

In cases of positive test results, employers' responses vary from workplace to workplace, as might be expected. According to one survey, "twenty-five percent of companies . . . automatically fire employees who use drugs" (Chineson 91). Others are less punitive. At many companies, a first offense will result in suspension and/or referral to a treatment program, with termination only for a second offense. From a legal standpoint, however, the worst case scenario is that of the employee who sues the company after being incorrectly terminated or referred; "even when employers . . . verify positive results, employees who turn out to be drug-free upon retesting will be . . . stigmatized" (O'Keefe 35). The consequences of such errors can be quite costly indeed. "The average jury award for wrongful termination of employment is $750,000 and the average legal costs are $125,000, win or lose" (Brookler 128–129). To guard against such outcomes, employers must design their drug testing programs very carefully.

V - CHARACTERISTICS OF AN
EFFECTIVE DRUG TESTING PROGRAM

Employers agree that for any drug testing program to succeed, every effort must be made to safeguard against error (and attendant liability) and to minimize any potentially negative impact upon employee morale. To this end,

FIGURE 10.1 **Drug Testing Report, page 6**

7

most progressive companies design their programs according to the guidelines published by the National Institute on Drug Abuse (NIDA). Some key features of a successful program are as follows:

- Clearcut Policy Statement

  Using input from human resources, employee relations, union and legal department representatives, and employees themselves, a clear, comprehensive policy statement must be written. The statement should spell out the company's standards of employee conduct, details of how and under what circumstances testing will occur, and what steps will be taken in response to a positive test result.

- Strict Guidelines for Specimen Collection

  A company may choose to collect specimens in-house (usually at the company's health or medical facility) or off-site, at a hospital or clinic, or at a facility specializing in such procedures. In any case, it is absolutely crucial that collection be conducted according to the strictest NIDA standards. "Because the first few links in the chain of custody are forged here . . . many experts feel that choosing your collection site merits greater attention than choosing your lab" (Brookler 130).

- NIDA-Certified Laboratory

  To become NIDA certified, a lab must meet the most stringent standards of accuracy and protocol, especially regarding the chain of custody governing the handling of specimens. In short, NIDA-certified laboratories are the most reliable and certainly the most persuasive in court.

- Confirmation of Positive Results

  Positive test results should be confirmed by means of a GC/MS follow-up test. "Without the GC/MS confirmation, you aren't legally defensible" (Brookler 129). In the event of a positive confirmation, the case must then be referred to the company's Medical Review Officer (MRO), a licensed physician knowledgeable about substance abuse, who searches for alternative medical explanations for the positive test result before providing final confirmation. The MRO may refer the case back to

FIGURE 10.1 Drug Testing Report, page 7

management only after the employee has been given the opportunity to meet with the MRO. (At most companies, the MRO is a contract employee.)

- Employee Assistance Program

  Since substance abuse is now recognized as a disease, many employers have begun to provide employee assistance programs. Rather than being terminated, employees who test positive may instead be referred for counseling. In such instances, the rehabilitation option is usually presented as a condition of continued employment.

CONCLUSION

Clearly, the whole subject of drug testing in the workplace is quite complicated. If Paragon Manufacturing, Inc. is to introduce a testing program, perhaps the way to begin would be to require testing of job *applicants* only—at least at first. This would enable us to become gradually acquainted with the procedures and problems involved, without the risk of alienating employees already on board. Perhaps we might expand the program at a later date, but not in the form of random testing, which has been shown to engender resentment and legal challenges. Any testing of the established workforce should be done only on a for-cause basis—in response to sudden absenteeism, erratic behavior, on-the-job accidents, and the like. And certainly we should establish provisions for an employee assistance program before any testing occurs. All of this, however, is by way of suggestion only. The next step should be to call in an outside consultant, preferably from NIDA, to provide additional input before any final decision is made.

FIGURE 10.1 **Drug Testing Report, page 8**

9

WORKS CITED

"An Alternative to Drug Testing?" *Inc.* April 1995: 112.

Brookler, Rob. "Industry Standards in Workplace Drug Testing." *Personnel Journal* April 1992: 128–132.

Chineson, Joel. "Mandatory Drug Testing: An Invasion of Privacy?" *Trial* Sept. 1986: 91–95.

Crow, Stephen M., and Sandra J. Hartman. "Drugs in the Workplace: The Problems and the Cures." *Journal of Drug Issues* 22 (1992): 923–937.

"Drug abuse at work." *The Economist* 30 Sept. 1989: 72.

Glantz, Leonard H. "A Nation of Suspects: Drug Testing and the Fourth Amendment." *American Journal of Public Health* 79. 10 (1989): 1427–1431.

Hukill, Craig. "Significant decisions in labor cases: Employee drug testing." *Monthly Labor Review* Nov. 1989: 75–77.

Konovsky, Mary A., and Russell Cropanzano. "Perceived Fairness of Employee Drug Testing as a Predictor of Employee Attitudes and Job Performance." *Journal of Applied Psychology* 76 (1991): 698–707.

Macdonald, Scott. "The Role of Drugs in Workplace Injuries: Is Drug Testing Appropriate?" *Journal of Drug Issues* 25 (1995): 703–722.

O'Keefe, Anne Marie. "The Case Against Drug Testing." *Psychology Today* June 1987: 34–38.

Payson, Martin F., and Philip B. Rosen. "Substance Abuse: A Crisis in the Workplace." *Trial* July 1987: 25–33.

West, Louis Jolyon, and Deborah L. Ackerman. "The Drug-Testing Controversy." *Journal of Drug Issues* 23 (1993): 579–595.

Zimmer, Lynn, and James B. Jacobs. "The Business of Drug Testing: Technological Innovation and Social Control." *Contemporary Drug Problems* 19. 1 (1992): 1–26.

FIGURE 10.1 Drug Testing Report, Works Cited

☑ Checklist Evaluating a Long Report

An effective long report

___ includes these components:

- ☐ transmittal document
- ☐ title page that includes student's name, instructor's name, course number/name/section, college, and date
- ☐ abstract that briefly summarizes the report
- ☐ table of contents, with sections numbered and titled and page numbers provided
- ☐ list of illustrations, each numbered and titled, with page numbers provided
- ☐ glossary, if necessary;

___ is divided into sections that are numbered and titled in conformity with the table of contents;

___ is well organized, adopting the most logical sequence for the information;

___ is clear, accurate, and sufficiently detailed to satisfy the needs of the intended audience;

___ uses clear, simple language;

___ maintains an appropriate tone, neither too formal nor too conversational;

___ employs tables, graphs, charts, and so on, each numbered and titled in conformity with the list of illustrations;

___ includes full documentation (bibliography and parenthetical citations) prepared according to MLA or APA format;

___ contains no typos or mechanical errors in spelling, capitalization, punctuation, and grammar.

Exercises

EXERCISE 10.1

Rewrite the transmittal memo and the Title Page in Figure 10.1, as if the report were your own work, submitted as an assignment in your Workplace Communications course.

EXERCISE 10.2

Rewrite the bibliography on page 225, using APA format.

EXERCISE 10.3

Consult the MLA Web site <http://www.mla.org>> to determine how to handle documentation for the following kinds of on-line sources:

- article in a reference database
- e-mail
- FTP (file transfer protocol), telnet, or gopher site
- personal Web site
- posting to a discussion list (listserv or newsgroup)
- synchronous communication (MOOs, MUDs, and IRCs)

■ EXERCISE 10.4

Guided by the table of contents that follows, team up with two or three other students to write a collaborative report entitled "Radar: History, Principles, Applications."

TABLE OF CONTENTS

■ EXERCISE 10.5

Practically all Workplace Communications courses include a long report assignment at some point during the semester, usually near the end. Specific features of the project, however, vary greatly from instructor to instructor. Write a long report designed to satisfy your own instructor's course requirements.

Appendix A

Ten Strategies to Improve Your Style

1. Create Active Sentences

In an active sentence, the subject is the actor, performing the action of the verb. In a passive sentence, the subject is *not* the actor, and in fact *receives* the action of the verb.

(Subject/Actor) (Verb/Action)

ACTIVE Dr. Polanski interviewed the patient.

(Subject/Receiver) (Verb/Action)

PASSIVE The patient was interviewed by Dr. Polanski.

Because the active approach requires fewer words, it is usually preferable. Of course, there are situations in which you may choose to use the passive approach—for emphasis or variety, perhaps. Or you may want to "hide" the actor, such as in writing about a situation in which some mistake or controversial decision has been made. Here's an example:

(Subject/Actor) (Verb/Action)

ACTIVE Ken Park decided that we should go on strike.

(Subject/Receiver) (Verb/Action)

PASSIVE A decision was made that we should go on strike.

Obviously, the passive voice would probably be better in a case like this. There is no reason to bring Ken Park's identity to the attention of anyone who may later wish to retaliate against strike organizers. But use the passive voice only with good cause; every use should be deliberate rather than accidental.

EXERCISE A.1

Revise these sentences by making them active rather than passive, but without changing the meaning.

1. Electrical shock can be avoided if certain basic rules are followed by the worker.
2. When the patient's parents were interviewed by Ms. Rodriguez, their responses were thought to be hostile.
3. In drilling holes for running cable through masonry, a half-inch, carbide-tipped masonry bit should be used by the contractor.
4. Low-wattage light bulbs should be purchased by the energy-conscious consumer.
5. All the necessary forms have been received by the social worker.
6. Preparation times for each dish should be known by the server.
7. When the patient's jacket was examined by the ward nurse, the buttons were seen to be missing.
8. If a lamp or ceiling fixture cannot be turned on by means of the switch, first the light bulb should be checked by the homeowner.
9. A dosage of 150 mg. of Imipramine is taken by this patient daily.
10. The knife blade is held flat against the sharpening stone, with the sharp edge facing right, and the blade is drawn across the stone from right to left. Then the blade is turned so that the sharp edge faces left, and the blade is drawn across from left to right. This procedure is repeated several times in rapid succession.

2. Keep Subjects and Verbs Close Together

The heart of any sentence is the subject and its verb. Whether active or passive, English sentences are easiest to understand when subjects and verbs are as close together as possible. For example, both of the following sentences are grammatical, but the second is smoother because the first creates an interruption between the subject and the verb.

(Subject)
Dr. Simmons, because she was concerned about safety,
(Verb)
recommended additional supervision of the pediatric ward.

(Subject) (Verb)
Dr. Simmons recommended additional supervision of the pediatric ward because she was concerned about safety.

Whenever possible, try to keep subjects and verbs side by side. This will not only ensure maximum clarity, but will also help you to avoid grammatical errors involving subject/verb agreement (see Appendix B, pages 294–295).

EXERCISE A.2

Rewrite these sentences, putting subjects and verbs closer together.

1. Because outdoor electrical conductors usually run underground, you must, if they are to endure years of burial in damp earth, enclose them in a conduit.
2. Wood doors, unlike steel, hollow-aluminum, or hollow-vinyl doors, react very little to heat and cold.
3. The construction of the new annex on Building 7, because inflation had not been taken into consideration when the project was given approval, is costing far more than originally anticipated.
4. In the case file, a complete report on the patient's past history of hospitalization is included.
5. The continuing deterioration of Willard's peer-group relationships, unless some form of alternative school placement is implemented, may eventually cause lasting psychological problems for him.
6. Welding, which requires manual dexterity, good hand-eye coordination, and a great deal of technical expertise, can be an art as well as a trade.
7. A transformer, by reducing the voltage and increasing the current, allows us to operate model trains on ordinary house current.
8. Silk-screen printing, as we know it today, is simply a more sophisticated version of the stencil printing techniques used in ancient times.
9. Food servers, because kitchen doors always open to the right, must learn to carry trays on the left hand.
10. Hans Christian Oersted, more than 170 years ago in Copenhagen University in Denmark, discovered that a current flowing steadily through a wire will set up a magnetic field around the wire.

3. Put Modifiers Next to What They Modify

A modifier is a word or a phrase that makes another word or phrase more specific by limiting or qualifying its sense. For example, in the phrase "the red car," the word "red" modifies the word "car." In the sentence "Exhausted, he fell asleep at the wheel," the word "Exhausted" modifies "he" and the phrase "at the wheel" modifies the word "asleep." And in the sentence "The social worker noticed marijuana growing in the closet," the word "growing" modifies the word "marijuana" and the phrase "in the closet" modifies the word "growing."

Obviously, then, it makes sense to put modifiers alongside what they modify, to avoid confusion or unintentionally comical effects. Notice what happens if the "social worker" sentence is misarranged:

> Growing in the closet, the social worker noticed marijuana.

Certainly the social worker was not "growing in the closet," yet that is what the sentence now says because the modifiers are in the wrong place, "dangling" off the front of the sentence.

Of course, nobody would be likely to misinterpret the sentence that way, but misplaced modifiers can indeed create misunderstanding when more than one interpretation is possible. Avoid problems by always putting modifiers next to what they modify.

■ EXERCISE A.3

Revise these sentences, putting the modifiers where they belong.

1. Sometimes it is profitable to buy an older home in a nice neighborhood that needs a lot of repairs.
2. Hector wrote his report at the last minute, while riding the bus to campus in the back of his notebook.
3. The word processor is now standard equipment in the office instead of the typewriter.
4. While welding a butt joint, several rules must be remembered.
5. Other matters were discussed in the closing moments of the meeting of lesser importance.
6. If the large geriatric population of the state hospital, with its inevitably high death rate, were not present, the overall hospital population would be growing rapidly.

7. In real French restaurants, all food is served from a cart that is kept close to the guests' table on wheels.
8. Inexpensive prints are available to art lovers in all sizes.
9. Placed in a window opening, office workers can be kept comfortable by an air conditioning unit of even moderate size.
10. The unopened wine bottle is presented to the guest, brought from the cellar, for approval before being placed in the cooler.

4. Adjust Long Sentences

A sentence can be very long and still be grammatical. Many highly regarded authors (the American novelist William Faulkner, for example) have favored elaborate sentence structure. But most of us are not literary artists. We write for a different purpose—to convey information, not to entertain, dazzle, or impress.

The more information you pack into a sentence, the harder it is for the reader to process. If a sentence goes much beyond twenty-five or thirty words, the reader's comprehension decreases.

To ensure maximum readability, therefore, limit your sentence length. Try for an *average* of no more than twenty words per sentence. Since this can be inhibiting during the actual composing process, make such adjustments afterward, when revising. Identify overly long sentences and break them down into shorter ones. It is usually fairly easy to see where the breaks should be, as in these examples:

ORIGINAL

Welding is a method of permanently joining two pieces of metal, usually by means of heat, which partially melts the surface and combines the two pieces into a single piece, and after the metal cools and hardens, the welded joint is as strong as any other part of the metal. (fifty words)

REVISION

Welding is a method of permanently joining two pieces of metal, usually by means of heat. The partially melted surfaces are combined into a single piece. After the metal cools and hardens, the welded joint is as strong as any other part of the metal. (sixteen words; ten words; nineteen words)

Another good strategy is to present lengthy information in the form of a list, as in this example:

ORIGINAL

There are four major methods of welding: arc welding, which joins metals by means of heat produced by an electric arc; resistance welding, which joins metals by means of heat generated by resistance to an electric current; gas welding, which joins metals by means of heat from a gas torch; and brazing, which joins metals by means of a melted filler metal such as brass, bronze, or a silver alloy.

REVISION

There are four major methods of welding:

- arc welding joins metals by means of heat produced by an electric arc;
- resistance welding joins metals by means of heat produced by resistance to the flow of an electric current;
- gas welding joins metals by means of heat from a gas torch;
- brazing joins metals by means of a filler metal such as brass, bronze, or a silver alloy.

Judging whether a sentence is too long, however, depends on its context: what comes before it and what comes after. An occasional long sentence is acceptable, especially before or after a short one.

EXERCISE A.4

Revise these sentences, breaking them down into shorter ones.

1. Electricians are usually expected to have their own hand tools, such as screwdrivers, pliers, levels, hammers, wrenches, and so forth, and an initial investment of several hundred dollars may be necessary to acquire an adequate supply of such tools, although the electrical contractor usually supplies large tools and pieces of equipment, such as hydraulic benders, power tools, ladders, and such expendable items as hacksaw blades, taps, and twist drills.
2. Used extensively and considered a general-purpose welding rod, RG-65 gas welding rods are of low-alloy composition and may be used to weld pipes for power plants, for process piping, and under severe service conditions, and produce very good welds (50,000 to 65,000 psi) in such materials as carbon steels, low-alloy steels, and wrought iron.

3. Fine silk-screen prints were first brought to public attention by Elizabeth McCausland, an art critic and writer who arranged and sponsored exhibits that astonished critics and laypersons alike, and of Carl Zigrosser, an art critic and writer, who arranged some of the earliest exhibitions of silk-screen print exhibitions.

4. Lack of fusion in welding may be caused by an incorrect current adjustment, an improper electrode size or type, dirty plate surfaces, failure to raise to the melting point the temperature of the base metal or the previously deposited weld metal, or improper fluxing, which fails to dissolve the oxide and other foreign material from surfaces to which the deposited metal is intended to fuse.

5. A headwaiter is responsible for the proper setup of tables and chairs before the dining room opens, and should therefore make a tour of the room beforehand to check on each setup and on the proper placement of furniture, and should also go over the menu with the wait staff, briefing them on the characteristics of menu items and on which items are ready to be served and which require longer preparation time.

6. Migration trends are moving the American population away from the Northeast and Midwest to the South and West, where land is more abundant and less expensive, and even within regions, people and jobs are moving away from the downtown areas to the suburbs and outlying areas, causing a housing market boom on the fringes of metropolitan areas, and a virtual halt to new construction within the inner cities.

7. Mobile equipment lasts longer and works better if operated on smooth floors by skilled operators, so floors should be kept clean, and (if possible) should be treated with urethane or other coatings to minimize dust, which contributes to wear on many lift truck parts, including the wheels, and operators' skills can be assessed by regularly inspecting the condition of the paint on the vehicles and checking doorways, building columns, and walls for damage indicating careless operation.

8. County jails have been the American penal institutions most resistive to change and reform, and even today are often unfit for human habitation, because characterized by unsanitary conditions, minimally qualified personnel, intermingling of all types of prisoners (sick and well, old and young, hardened criminals and petty offenders) in overcrowded cellblocks and "tanks," and

the almost complete absence of even the most rudimentary rehabilitative programs, constituting a scandalous state of affairs that will be eliminated only when the public begins to support the many sheriffs and jailers who are trying to correct bad conditions and practices.

9. North American rodents include such native animals as field and wood mice, wood rats, squirrels, rabbits, woodchucks, gophers, muskrats, porcupines, and beavers, along with three more destructive species—the house mouse, the roof rat, and the Norway rat—that reached the United States from other countries and can be found everywhere, inhabiting buildings and eating or damaging stored food.

10. A wholesale travel agent may design tour packages marketed under the agency's name, or may take land packages already assembled by a ground operator and combine them with air or surface transportation to form new packages, an especially common practice in the international tourism field where, for example, rather than negotiating directly with hotels, sightseeing operators, and the like, a German wholesaler wishing to offer a New York City program might contract with a New York ground operator for land arrangements and then add international air transportation between the point of origin and the destination.

5. Use Transitions

Transitional words or phrases can serve as links between sentence parts, whole sentences, or paragraphs, clarifying the direction of your train of thought. In effect, they serve as "bridges" from idea to idea within a piece of writing, and are therefore quite helpful to your reader. Usually positioned at the beginning of a sentence and followed by a comma, these devices can be loosely categorized according to the various relationships they signal. Here are some common examples:

Additional Information: also, furthermore, in addition, moreover

Exemplification: for example, for instance, in other words, to illustrate, specifically

Explanation: in other words, put another way, simply stated

Similarity: in like manner, likewise, similarly

Contrast: conversely, however, nevertheless, on the contrary, on the other hand, yet

Cause and Effect: accordingly, as a result, consequently, hence, therefore, thus

Emphasis: clearly, indeed, in fact

Summary: finally, in conclusion, in short, to sum up

Helpful as they are, transition devices should not be used excessively. Employ them selectively, and only where they will provide needed assistance. If the relationship between the ideas is already clear, no transition is needed.

Always exercise great care in selecting transitions. Since these devices greatly influence how a sentence is interpreted, an incorrect choice will badly obscure your meaning. Consider this example.

> Wilson Brothers submitted a bid of $10,000 for the project. Therefore, we hired them.

In this case, the transition "Therefore" indicates that the bid was quite competitive. But a different transition, as in this second example, might indicate just the opposite.

> Wilson Brothers submitted a bid of $10,000 for the project. Nevertheless, we hired them.

Writers should always choose their words carefully; developing your skill in selecting appropriate transition devices is an important application of that principle.

■ EXERCISE A.5

Improve each of these sentence pairs by starting the second sentence with a transition device that clarifies or reinforces the meaning.

1. The company's engineering department has proudly announced that the number of new projects has more than doubled during the past year. The accounting department has repeatedly asserted that our main focus should not be on the number of new projects but on their relative cost.

2. Area businesses seeking to attract new employees to the region are disturbed by rising crime rates. Armed robbery has increased 12 percent locally since 1997.
3. As forecast last spring, corporate donations to the Foundation now total more than a million dollars. Private donations have also met our expectations.
4. Many factors can lower employee morale. Unsafe working conditions are a major source of discontent.
5. For the company to remain viable, we must generate more revenue than expenditures, realizing enough gain to repay our investors. We must achieve a positive profit margin.
6. The motion sensors had somehow been deactivated. The intruders were able to enter the restricted area undetected.
7. This is a major obstacle to continued growth and product development. It is the single biggest problem facing the company.
8. Within the next six months we must hire three welders, four electricians, and two carpenters. We will need at least two new clerical workers.
9. We have finished the project within budget, on time, and to everyone's complete satisfaction. We have succeeded.
10. Management has offered the salaried employees a non-retroactive 2 percent raise. The union is demanding a retroactive 3 percent increase.

6. Eliminate Clutter

Practically everyone wastes words, at least in a first draft. But excess verbiage interferes with communication by inflating sentence length and tiring the reader. Therefore, try not to use any more words than necessary. For example, do not say

> The attitude of the supervisory individual is a good one at the present time. (fourteen words)

Instead, say

> The supervisor has a good attitude now. (seven words—*half* as long)

There are many kinds of verbal clutter. This section discusses the five most common.

Unnecessary Introductions

It is perfectly acceptable—and sometimes necessary—to open a sentence with an introductory phrase that leads into the main idea. But this depends on who the reader is, what the circumstances are, and other factors. Check your writing for *needless* introductions, phrases in which you are simply spinning your wheels, as in these examples:

> As I look back on what I have said in this memo, it seems as if . . . .
>
> Although this may not be very important, I think . . . .
>
> Since all of us attended last week's meeting, there is no need to summarize that discussion, in which we agreed to . . . .

Instead, get right to the point:

> It seems as if . . . .
>
> I think . . . .
>
> At last week's meeting we agreed to . . . .

Submerged Verbs

Too often, we use a verb plus another verb (hidden or "submerged" within a noun) when one verb (the "submerged" one) would do, as in phrases like the following:

(Verb) (Noun) (Verb)
conduct an investigation = investigate

(Verb) (Noun) (Verb)
reach a decision = decide

(Verb) (Noun) (Verb)
give a summary = summarize

Instead, simply use the "submerged" verb: "investigate," "decide," "summarize." This approach is far better because it is more direct.

Long-Winded Phrases

Submerged verbs are one source of clutter, but any long-winded expression using more words than necessary wastes the reader's time and energy. Here are ten common examples, with suggested revisions:

at this point in time	=	now
despite the fact that	=	although
due to the fact that	=	because
during the time that	=	while
in many instances	=	often
in order to	=	to
in the course of	=	during
in the event that	=	if
in the near future	=	soon
on two occasions	=	twice

Obvious Modifiers

Sometimes we use words unnecessarily, expressing already self-evident ideas, as in these modifiers:

(Modifier) personal opinion	obviously, *all* opinion is "personal"
(Modifier) visible to the eye	obviously, anything visible *must* be visible "to the eye"
(Modifier) past history	obviously, *all* history is "past"

In these cases, it would be better to omit the modifiers and simply say "opinion," "visible," and "history."

Repetitious Wording

As we have seen, several short sentences are usually better than one long one. But sometimes it is better to *combine* two or three short sentences to avoid unnecessary repetition. If done correctly, the resulting sentence will often be significantly shorter than the total length of the several sentences that went into it, as in this example:

ORIGINAL

The electric drill is easier to use than the hand drill. The electric drill is faster than the hand drill. The electric drill is a very useful tool. (twenty-eight words)

REVISION

Easier and faster to use than the hand drill, the electric drill is a very useful tool. (seventeen words)

In the original, "the electric drill" appears three times and "the hand drill" twice. By using each of these phrases only once, the revision conveys the same information much more efficiently. Always strive for this level of economy.

EXERCISE A.6

Revise these sentences to express the ideas more concisely.

1. The company cannot maintain full employment during the summer months.
2. In the event that the conductor becomes hot, shut down the unit.
3. I sent a check in the amount of $379 on April 3.
4. This airbrush has a tendency to leak.
5. The Security Office should take into consideration the feelings of the other employees.
6. Fire drills are important because fire drills provide students with practice in emergency evacuation procedures that they will have to know in the event of an actual fire.
7. On all our computers, the "Delete" key is red in color.
8. The Desert Storm war was relatively brief in duration.
9. In view of the fact that Christmas falls on a Monday this year, we shall have a long weekend.
10. Prior to entering the factory, please sign the visitors' log book.

7. Choose Simple Language

Most readers do not respond well to fancy or unnecessarily complex wording. Therefore, you should instead use ordinary terms your reader will immediately recognize and understand.

Of course, there is no reason to avoid *technical* terms—specialized words for which there are no satisfactory substitutes—if your reader can be expected to know them. An electrode is an electrode, a condenser is a condenser, and an isometric drawing is an isometric drawing. If you are writing in a technical context, your reader often will share your knowledge of the subject area. If you are not sure whether the reader is familiar with the vocabulary of the field, provide a glossary (a list of words with their definitions), as discussed in Chapter 10. If there are only two or three questionable terms, you can insert parentheses—as in the previous sentence, in which "glossary" is defined.

Generally, however, the best policy is always to use the simplest word available: "pay" rather than "remunerate," "transparent" rather than "pellucid," "steal" rather than "pilfer." Another advantage of using everyday vocabulary is that your spelling will improve. You are far more likely to misspell if using words you are unaccustomed to seeing in print, because you will not know whether they "look right" on the page.

■ EXERCISE A.7

Revise these sentences by replacing the underlined words with simpler ones. You may have to consult a dictionary to determine what some of the underlined words mean. (Of course, that is why they are poor choices.)

1. These guidelines cover everything <u>germane</u> to the procedure.
2. In February we shall <u>commence</u> to use the new equipment.
3. The worst <u>scenario</u> would be if the company had to relocate.
4. Supervisors should <u>transmit</u> accident reports to the personnel office.
5. The summer schedule will <u>terminate</u> on September 10.
6. Please <u>utilize</u> safety glasses when working with the grinding machines.
7. All workers should <u>eschew</u> procedures that are obviously dangerous.

8. You must complete the <u>requisite</u> forms before we can consider your request for personal leave.
9. The union will <u>endeavor</u> to avoid a strike, but the company must meet certain conditions.
10. Let me explain what <u>transpired</u> at last Thursday's meeting.

8. Avoid Clichés

Sometimes a catchy or cleverly worded phrase will quickly gain widespread popularity and become a permanent part of a language. Ironically, however, the phrase then loses much of its original energy, becoming stale and lifeless through overuse. It becomes, in short, a cliché. Countless clichés exist in English (as in most languages) and their number has increased greatly during the past couple of decades, due to the growing influence of advertising and the mass media.

Indeed, clichés have become so much a part of the culture that many people rely heavily on them when speaking. This is hardly surprising, since such phrases require no real thought and are readily understood. But clichés should be avoided in writing. Because they are by nature "prefabricated," they create an aura of trite predictability that robs language of vitality. In addition, clichés are often wordy. For example, why say "at this point in time" or "easier said than done" when "now" or "difficult" would suffice?

Here are ten more very common clichés, with suggested alternatives.

between a rock and a hard place	in a difficult situation
beyond the shadow of a doubt	undoubtedly
few and far between	scarce
first and foremost	first
in the final analysis	finally
let bygones be bygones	forgive
needless to say	obviously
a rude awakening	a shock
see eye to eye	agree
tried and true	proven

EXERCISE A.8

Revise these sentences by replacing the underlined clichés with more original wording.

1. When the company switched insurance carriers we went from the frying pan into the fire.
2. For all intents and purposes, the agency is now performing the duties of two or three agencies.
3. The heart of the matter is that the employee parking lot is too small.
4. If we could expand our product line, this would be a whole new ball game.
5. Certainly it would be a step in the right direction to expand the company's marketing base.
6. The outdated machinery sticks out like a sore thumb.
7. It is time for the company to throw caution to the wind and make a major financial commitment to expansion.
8. If profits continue to decline, somebody will have to face the music.
9. If we play our cards right we will be able to survive this crisis still smelling like a rose.
10. Last but not least, we must consider the possibility of relocating the downtown office.

9. Use Numbers Correctly

Since much workplace communication involves quantitative information, writers often must decide whether to spell out a number or use a numeral (e.g., "three hundred" vs. "300"). This issue is somewhat complicated, because several different systems exist. In general, workplace writers prefer numerals to words. Often, however, words are used from zero to nine, and sometimes for other amounts that can be expressed in one or two words, such as "thirty-one" or "fifty." (Note that two-word numbers under 100 require a hyphen.)

Here are a few guidelines that may help.

- Never begin a sentence with a numeral. Either spell it out or, if the number is large, reorder the sentence so that the numeral appears elsewhere.

  ORIGINAL 781 people have registered for the convention so far.

  REVISION So far 781 people have registered for the convention.

- For very large numbers, combine numerals and words, as in "100 million."
- Combine numerals and words whenever such an approach will prevent misreading, as in this example:

  ORIGINAL You will need 3 6 inch screws and 10 4 inch nails.

  REVISION You will need three 6-inch screws and ten 4-inch nails.

  Note that a hyphen is required when a number is used with a word to modify another word.

- Be consistent about how you deal with numbers in a given piece of writing. Pick one approach and stick with it, as in this example:

  ORIGINAL We will hire 5 masons and two plumbers.

  REVISION We will hire five masons and two plumbers.

- Use numerals for all *statistical* data, such as the following:

  ages and addresses
  dates
  exact amounts of money
  fractions and decimals
  identification numbers
  measurements (including height and weight)
  page numbers
  percentages, ratios, proportions
  scores

  When dealing with statistics, it often makes sense to combine numerals with other symbols rather than with words (e.g., 6'1" rather than 6 feet, 1 inch and $375 rather than 375 dollars).

Always review your completed draft to ensure that you have handled numbers in a correct, consistent manner.

EXERCISE A.9

Revise these sentences to correct all errors involving numbers.

1. A test of thirty one control valves at a major factory revealed more than seventy significant operating deficiencies, and in another series of tests conducted on a random sample of 60 control valves, 88 percent exhibited substandard performance.
2. $15,140 in "play money" is included in every Monopoly game.
3. This machine offers a calculated mean time between failures (MTBF) above one hundred thousand hours, or more than eleven years if used twenty-four hours per day, three hundred and sixty-five days per year.
4. The job will involve undersealing, lifting, and stabilizing a fifteen thousand square foot concrete slab.
5. In nineteen ninety three, a one hundred and eighty foot long, three hundred and twenty eight ton replacement span for the historic Grosse Island Bridge in Michigan was set in place in one piece, using the "sinking bridge" method.
6. In the nineteen seventies, baseball's Major Leagues comprised 4 divisions—the American League East, the American League West, the National League East, and the National League West—each with 6 teams.
7. The population of the United States is now more than two hundred fifty million.
8. It is unusual for a high school basketball team to have more than 1 or 2 20 point scorers.
9. The temperature in Death Valley often exceeds one hundred degrees.
10. When it is six o'clock in New York, it is 3:00 o'clock in California.

10. Revise Sexist Language

In recent years we have begun to understand that English (like most other languages) tends to be male-oriented, as is the traditional view of society itself. Now, however, men's and women's social roles are changing. Women are working in jobs from which they would have been excluded only a short time ago, and in general we are evolving toward a more sophisticated, less restrictive sense of the relationship between the sexes.

Certainly we should try, therefore, to bring our language usage up to speed. After all, language not only reflects social values, but also reinforces them. Clinging to old-fashioned constructions only perpetuates outdated thinking.

Here are four examples of sexist (i.e., gender-biased) writing:

> Every welder must sign his name on the log-in sheet before beginning his shift.
>
> Mr. Jones, Miss Gomez, and Mrs. Ching will be assigned to the budget committee.
>
> Lollipop Nursery School requests that each child's mother help out at the school at least one lunch hour per month.
>
> It will take twelve workmen to complete this job on time.

Although all of these sentences are grammatically correct, each is sexist. The "Every welder" sentence, by twice using the word "his," implies that all welders are male—certainly not true today! But the sexism can easily be removed, just by cutting a few words ("his name on") and changing "his shift" to "work":

> Every welder must sign the log-in sheet before beginning work.

In the "Mr. Jones" example, we can assume that Ching is married and Gomez is single. But Jones's marital status remains undetermined, as is appropriate; such matters have no bearing on one's role in the workplace. Equal consideration should be granted to all three employees, by referring to both women as "Ms.," or by dropping the titles altogether. In addition, the three names should appear in alphabetical order, as there is no apparent reason why the male name should automatically stand first:

> Ms. Ching, Mr. Jones, and Ms. Gomez will be assigned to the budget committee.

or

> Ching, Jones, and Gomez will be assigned to the budget committee.

The writer of the "Lollipop" sentence implies that childcare is solely the responsibility of children's mothers—hardly the case! A far better phrasing would be to replace "mother" with "parent(s)":

> Lollipop Nursery School requests that each child's parent(s) help out at the school at least one lunch hour per month.

The "twelve workmen" sentence implies that only males (work*men*) could do this job. Avoid gender-biased terms like "workman"—or "fireman," "mailman," and "policeman"—and instead use gender-neutral ones like "worker"—or "firefighter," "mail carrier," and "police officer":

> It will take twelve workers to complete this job on time.

In the interest of simple fairness, we must all develop the habit of non-sexist expression!

■ EXERCISE A.10

Revise these sentences to eliminate sexist language.

1. To keep his knowledge up to date, a doctor should read the leading medical journals in his field.
2. Every man needs the satisfaction provided by meaningful work.
3. The average employee is dissatisfied with his wages.
4. As we move through the twenty-first century, every American president will have to work very closely with his advisors to prevent the occurrence of another major war.
5. There will be a policeman posted at each of the entrances, and nobody will be allowed into the plant without his ID card.
6. I now pronounce you man and wife.
7. In the early days of public education, a schoolboy's textbooks had to be purchased by his parents.
8. Increasingly, manufacturers are using man-made materials rather than natural ones.
9. If a patient decides to leave the hospital against medical advice, he must first sign a waiver of hospital responsibility.
10. Throughout history, man has created many remarkable inventions.

Appendix B

Review of Mechanics: Spelling, Punctuation, and Grammar

Spelling

Most writers experience at least some difficulty with spelling. You can become a better speller, however, simply by observing the following basic guidelines.

1. Do not concern yourself with spelling while you are composing. Concentrate on content instead. But at the re-writing stage, check carefully for obvious errors—words that you know how to spell but got wrong through carelessness. Do not permit obvious blunders to slip past you. When in doubt, consult the dictionary or an electronic spelling checker. Most electric typewriters and virtually all word processors are equipped with such devices. These are not foolproof, however. Consider this little poem that was making the rounds a few years ago. It illustrates quite well (if somewhat jokingly) the shortcomings of spelling checkers.

   I have a spelling checker.
   It came with my p.c.
   It plainly marks **four** my review
   mistakes that I **cant sea.**
   I **rote** this poetry **threw** it,
   I'm **shore your** pleased **too no.**
   Its letter perfect in **it's weigh.**
   My checker **tolled** me **sew.**

   Of the forty-four words in this small poem, fourteen (the ones in bold print) are incorrect—an error rate of nearly 30 percent! Yet no standard spelling checker would identify these mistakes, since all are in fact actual words. Although quite helpful in spotting typographical miscues and other such flaws, spell checkers are no substitute for vigilance on the part of the writer.

2. Certain pairs of homonyms—words that sound alike but are spelled differently—give nearly everyone trouble. Memorize this list of commonly confused words:

   accept: to receive willingly
   except: with the exception of

   affect: to produce an effect upon
   effect: that which is produced, a result

   alot: [no such word]
   a lot: a great many

   a while: a period of time
   awhile: for a period of time

   cite: to quote or mention
   sight: something seen
   site: the position or location of something

   its: possessive form of **it**
   it's: contraction meaning **it is**

   loose: not tight
   lose: to misplace

   passed: past tense of **to pass**
   past: gone by in time

   their: possessive form of **they**
   there: in or at that place
   they're: contraction meaning **they are**

   to: a word used to express movement toward
   too: also, more than enough
   two: 2

   whose: possessive form of **who**
   who's: contraction meaning **who is** or **who has**

   your: possessive form of **you**
   you're: contraction meaning **you are**

3. We all have certain words we nearly always misspell, a handful of terms we repeatedly get wrong. Identify your own "problem" words, make a list of them, and consult it whenever you must use one of those words. Eventually, you will no longer need the list, as the correct spellings imprint themselves on your memory.

4. Stay away from fancy terms you are unaccustomed to seeing in print. Use ordinary, everyday words instead. Not only will your reader understand them more easily, but also you will be more likely to notice if they "look wrong" because of misspelling.

5. For the spelling of specialized or technical terms, check manuals and the indexes of textbooks in your field of major. For the spelling of the names of local persons, businesses, addresses, and so forth, consult the telephone directory. (The Area Code section even lists all states and major cities.) Most businesses and organizations provide in-house directories of employees; frequently these are available via on-line database.

6. Memorize some basic rules. Even though English spelling is highly inconsistent and filled with exceptions, there are some generally reliable patterns that can be learned, as follows.

   - "i" before "e," as in

     achieve, believe, retrieve

     except after "c," as in

     receive, conceive, deceive

     Note that in all the above examples the two letters "**i**" and "**e**" combine to sound like a "long e." If they combine to sound like anything else, the "i" before "e" except after "c" rule no longer applies, as in:

     height ["long i" sound] and **weight** ["long a" sound]

   - When adding a suffix (an "ending") to a word that ends in "e," keep the "e" if the suffix begins with a consonant, as in:

     complete completely;
     require requirement;
     wire wireless

     Drop the "e" if the suffix begins with a vowel, as in:

     electrocute electrocution;
     wire wiring

   - When adding a suffix to a word that ends in a consonant followed by "y," change the "y" to "i" unless the suffix begins with "i," as in:

     photocopy photocopier;
     hurry hurried;
     fry frying

When adding a suffix to a word that ends in a vowel preceded by "y," keep the "y," as in:

dye dyes

- When adding a suffix to a word that ends in a consonant, double the consonant only if
—the consonant is preceded by a vowel;
—the word is one syllable, or accented on the last syllable;
—the suffix begins with a vowel. Example:

  begin beginning

- When choosing between the suffixes -able and -ible, remember that most of the -able words are "able" to stand alone without the suffix, as in **laughable** and **paintable**, while most of the -ible words cannot, such as **horrible** and **responsible.**

- To make a word plural, add -es if the pluralizing creates an extra syllable, otherwise just add an -s.

  1 switch, 2 switches;
  1 hammer, 2 hammers

  If the word ends in a vowel followed by "o," add -s.

  1 radio, 2 radios;
  1 video, 2 videos

  If the word ends in a consonant followed by "o," add -es, as in:

  1 tomato, 2 tomatoes;
  1 potato, 2 potatoes

  Remember never to pluralize a word by using an apostrophe. The apostrophe is used only in contractions and to indicate possession. (See pages 288–289.)

EXERCISE B.1

There is at least one misspelled word in each of the following sentences. Rewrite the sentences to correct all spelling errors.

1. Specialized method's must be used for recycling hazardous waist products such as batterys, paint, and used oil.
2. Although bridges often go unoticed, there function is esential to modern America.
3. The principle component's of natural dry gas are methane, ethane, and propayne.
4. The viscosity of oil varys, but oil is allways lighter then water.

5. Biologists have mounted tiny bar codes on bees to moniter the insects' mateing habbits.
6. In the snowbelt states, highway agencys have been experamenting with new kind's of salt products during the passed severel years.
7. Obviousley, not all technological problems are easily solveable.
8. Its extremely dangerous to allow an open flame anywhere near a combustable gas.
9. American moterists have been slow to except the idea of anything but a gasaline-powered car.
10. When shifting to a lower geer, a vehikle gains power but looses speed.

Punctuation

Punctuation exists not to make writers' lives more difficult, but to make readers' lives easier. A punctuation mark is simply a symbol, like a road sign on a highway. It tells readers when to slow down, when to stop, and how to anticipate and respond to what appears before them on the page. This brings us to a basic principle: *Trust your ear; listen to the sentence and insert punctuation marks wherever you can hear them.*

Be careful, however, not to over-punctuate. Just like on a highway, an incorrect sign is even more misleading than a missing one. Consider this sentence, for example:

> The Carrier Corporation manufactures a wide variety of air conditioning units, ranging from small household models to huge industrial installations.

If the comma were removed, no real harm would result, because the reader would probably still pause after "units" instinctively, just to draw breath. And certainly the removal of the comma would in no way change or hide the meaning. But look at this version:

> The Carrier, Corporation manufactures, a wide, variety of air conditioning units, ranging from, small, household models, to huge, industrial installations.

See how much harder reading the sentence has become? All those unnecessary commas cause the reader to hesitate repeatedly, thereby derailing the train of thought. And the reader will keep pausing, because we automatically respond to symbols, whether we want to or not. This leads to a second basic principle: *Do not punctuate at all unless you are absolutely sure you must; when in doubt, leave it out.*

Of course, there is more to punctuation than these two principles. As you know, there are literally hundreds of punctuation rules. But the good news is that—unless you plan to become a professional writer or editor—you need to know only a small percentage of them. This section of Appendix B will teach you the basics of punctuation, the fundamental rules everyone needs to know in order to write clearly.

"End" Punctuation: The Period, the Question Mark, and the Exclamation Mark

Practically everyone knows how to use periods, question marks, and exclamation marks at the end of sentences. Sometimes, however, we simply forget to use end punctuation because the mind is faster than the hand. As we write, we tend to think ahead a sentence or two, and it is easy to overlook the punctuation in our rush to express the next thought. This is something to watch for at the rewriting stage. Make sure that every sentence ends with punctuation. Be especially vigilant about question marks. A common error is to hastily insert a period even though the sentence is actually a question. As for exclamation marks, use them rarely. If they become commonplace, they lose their emphasis and hence their impact. Use exclamation marks only when really necessary, to signal an unusually emphatic statement.

The Comma

The comma is probably the most difficult of all punctuation marks to use correctly, because it is required in such a wide range of situations. If you study the following comma rules, however, you will notice that it is also the easiest punctuation mark to "hear." As mentioned earlier, trust your ear for determining where to insert the comma. And when in doubt, leave it out. The following rules will help guide you.

1. Use a comma to separate words or phrases in a sequence, as in these examples:

   The most common crimes are murder, rape, robbery, assault, burglary, and theft.

   There are five main types of metering devices used in refrigeration: the automatic expansion valve, the thermostatic expansion valve, the capillary tube, the low-side float, and the high-side float.

Notice that there is a comma before the "and" in both of these sequences. Some writers choose to skip that last comma, but it is better to use it.

2. Use a comma after an introductory word or phrase, as in these examples:

    Obviously, safety is an important consideration in the workplace.

    Because of its flexibility, a drive belt can be used to connect non-parallel shafts.

    In a typical transistor, the collector-base circuit is reverse-biased.

3. Use a comma *before* "linking" words such as "and," "but," or "so," if the word is linking two complete sentences, as in these examples:

    The upper number in a fraction is called the numerator, and the lower number is called the denominator.

    The risks involved in underwater welding are great, but the pay is excellent.

    Pressure within a fluid is proportional to the density of the fluid, so gasoline exerts less pressure than water.

4. Use commas between two or more adjectives in a row, but only if those adjectives would make sense in any order, as in this example:

    This task requires a large, heavy sledge.

    But in the following example, there is no comma between the adjectives ("three," "more," and "full") because they make sense only in the order given:

    Give the valve handle three more full turns.

5. Use commas to surround words or phrases that are not 100 percent essential to the sense of the sentence—words or phrases that could just as easily appear in parentheses, as in these examples:

    The catapult, a device for hurling large objects, was an early war machine.

    Pavlov, a Russian physiologist, demonstrated the existence of conditioned reflexes by conditioning dogs to salivate at the sound of a bell.

    The Louvre, located in Paris, is a major tourist attraction.

6. Use a comma before an afterthought, a word or phrase "tacked on" to the end of a sentence, as in these examples:

   Lead is highly toxic and extremely dangerous, even more so than previously believed.

   The first really good synthetic abrasive was carborundum, introduced in 1891.

   Drunken drivers cause many accidents, often after falling asleep at the wheel.

The Colon and the Semicolon

The colon and the semicolon are obviously related, but they serve different purposes and should not be used interchangeably.

The four main uses of the colon are after the salutation of a business letter and after the headings of a memo (see Chapter 2), to introduce a complicated list, and to introduce a long quote, as in these examples:

> At an 1874 meeting in Switzerland, delegates from twenty-two countries decided on four main principles that still govern postal service today: the postal service of the world should be regulated by a common treaty; the right of transit by sea or land should be guaranteed by every country to every other country; the country of origin should be responsible for the transmission of mail, and all the intermediate services should be paid for by fixed rates and according to periodic statistics; and each country should keep all its postage collections on both paid and unpaid letters.

> All American school children are required to learn the Pledge of Allegiance: "I pledge allegiance to the flag of the United States of America and to the Republic for which it stands, one Nation under God, indivisible, with liberty and justice for all."

Other uses of the colon are in denoting time of day (3:05 a.m.) and "stopwatch" time (1:07:31), in Biblical citations (Corinthians 3:22), in two-part book and article titles (*Workplace Communications: The Basics*), and in various locations within bibliography entries (see Chapter 10). It can also be used to serve a "stop/go" function, as in "Some people are motivated by only one thing: money."

There are really only two uses for the semicolon: to link two complete sentences that are closely related, and to separate the items in a complicated list, as in the "postal service" example.

The semicolon should be avoided, however, because there is usually a better alternative. When linking closely related sentences, for example, a comma along with a linking word like "and," "but," or "so" will not only establish the connection but also clarify the relationship between the ideas. Compare these examples:

> He ate too much; he became ill.
>
> He ate too much, so he became ill.

Clearly, the second version is better. It not only connects the two sentences, but plainly shows that the second idea is a result of the first.

As for separating the items in a complicated list, simply arrange the items vertically, like this:

> At an 1874 meeting in Switzerland, delegates from twenty-two countries decided on four main principles that still govern postal service today:
>
> - The postal service of the world should be regulated by a common treaty.
> - The right of transit by sea or land should be guaranteed by every country to every other country.
> - The country of origin should be responsible for the transmission of mail, and all the intermediate services should be paid for by fixed rates and according to periodic statistics.
> - Each country should keep all its postage collections on both paid and unpaid letters.

Again, the second version is obviously preferable. It enables the reader to differentiate better among the separate items.

Quotation Marks

Quotation marks are used to surround a direct quotation (someone else's exact words); to isolate a word within a sentence; or to show that a word is being used sarcastically, ironically, or in some other non-literal way. Here is an example of each such application:

DIRECT QUOTE	As Thomas Edison said, "There is no substitute for hard work."
ISOLATED WORD	In the above sentence, "said" is the fourth word.
NON-LITERAL USE	An athlete must not "choke" under pressure.

Quotation marks must be positioned correctly in relation to other punctuation, especially end punctuation. Follow these guidelines:

- A period at the end of a sentence always goes inside the quotation marks.

  Caesar said, "I came, I saw, I conquered."

- A question mark at the end of a sentence goes inside the quotation marks if the quote itself is a question.

  Juliet cries, "Wherefore art thou Romeo?"

- A question mark at the end of a sentence goes outside the quotation marks if the whole sentence (rather than the quote) is a question.

  Did Caesar say, "I came, I saw, I conquered"?

- A question mark at the end of a sentence goes inside the quotation marks if the quote and the whole sentence are both questions.

  Does Juliet cry, "Wherefore art thou Romeo?"

- If an attributing phrase follows the quote, the comma goes inside the quotation marks.

  "I came, I saw, I conquered," said Caesar.

- If an attributing phrase follows a quote that is a question, omit the comma but leave the question mark inside the quotation marks.

  "Wherefore art thou Romeo?" cries Juliet.

The Apostrophe

The apostrophe is often misused. In actuality, however, the rules governing this punctuation mark's use are quite simple.

1. NEVER use the apostrophe to make a word plural.

   INCORRECT Carpenter's use a variety of tool's.

   CORRECT Carpenters use a variety of tools.

2. Use the apostrophe to make a word possessive, as follows:

   - If the word is singular, add 's

     one man's hat
     one woman's hat
     John Jones's hat
     Jane Smith's hat

- If the word is plural and does not already end in -s, add 's

  the children's hats
  the men's hats
  the women's hats

- If the word is plural and already ends in -s, add an apostrophe

  the boys' hats
  the girls' hats
  the Joneses' house
  the Smiths' house

3. Use the apostrophe to replace the missing letter(s) in a contraction.

I am = I'm	I will = I'll
should have = should've	could have = could've
would have = would've	it is = it's
she is = she's	she will = she'll

EXERCISE B.2

Punctuation has been omitted in each of these paragraphs. Rewrite the sentences, dividing them into new sentences if appropriate, and inserting end punctuation and other punctuation wherever necessary.

1. Refrigerants are used in the process of refrigeration whereby heat is removed from a substance or a space a refrigerant is a substance that picks up latent heat when the substance evaporates from a liquid to a gas this is done at a low temperature and pressure a refrigerant expels latent heat when it condenses from a gas to a liquid at a high pressure and temperature the refrigerant cools by absorbing heat in one place and discharging it in another
2. Full-mold casting is a process in which a green sand or cold-setting resin bonded sand is packed around a foamed plastic pattern (for example polystyrene) the plastic pattern is vaporized when the molten metal is poured into the mold an improved casting surface can be obtained by putting a refractory type of coating on the pattern surface before sand packing the pattern can be one piece or several pieces depending on the complexity of the part to be cast
3. The first step in bias circuit design is to determine the characteristics of both the circuit and the transistor to be used will the cir-

cuit be used as an amplifier oscillator or switch what class of operation (A, AB, B, or C) is required how much gain (if any) is required what power supply voltages are available what transistor is to be used must input or output impedances be set at an arbitrary value

■ EXERCISE B.3

Punctuation has been omitted in each of these paragraphs. Rewrite the sentences, inserting commas and other punctuation wherever necessary.

1. If the Bonneville Salt Flats have meant one thing to nature lovers to enthusiasts for speed their meaning has been something else again. There has been no finer natural speedway on earth or at least none so readily accessible. The water table generally about a foot down rises to the surface during the rainy season and water collects to a depth of a foot or more. This water is gently swirled around by the wind and by the time it evaporates the surface is left smooth level and almost as hard as concrete.
2. Emotional players have no place on a Japanese baseball team. Those who get into fights in the clubhouse or enjoy practical jokes may be relieving tension on an American team but are only contributing to it on a Japanese team. The good Japanese team is composed of players who never argue never complain and never criticize each other. The good team is like a beautiful Japanese garden. Every tree every rock every blade of grass has its place. When each player's ego detaches itself and joins twenty-five others to become one giant ego something magical happens. All the efforts and sacrifices the players have made at last become worthwhile for they are now a perfectly functioning unit.
3. Several avenues are open to baseball card collectors. Cards can be purchased in the traditional way at the local candy grocery or drug stores with the bubble gum or other products included. For many years it has been possible to purchase complete sets of cards through mail order advertisers found in traditional sports media publications such as magazines newspapers yearbooks and others. Sets are also advertised in the card collecting periodicals. Many collectors begin by subscribing to at least one of the hobby periodicals all with good up-to-date information. In addition a great variety of cards can be obtained at the growing number of hobby retail stores around the country.

EXERCISE B.4

Consult some newspapers and magazines and find ten examples of sentences in which the colon introduces a list, and ten examples in which the colon introduces a quote.

EXERCISE B.5

Refer to Exercise B.4 and copy the sentences in which the colons introduce lists. Then rewrite each sentence, removing the semicolons and arranging the entries as a vertical list.

EXERCISE B.6

Again consulting newspapers and magazines, find ten examples of semicolons linking complete sentences. Rewrite those sentences, replacing each semicolon with a comma and a linking word.

EXERCISE B.7

Consult some newspapers and magazines and copy ten sentences in which quotation marks appear. Have the quotation marks and surrounding punctuation been correctly positioned in relation to each other? Rewrite any that are incorrect.

EXERCISE B.8

Consult some newspapers and magazines and copy ten sentences in which the apostrophe appears. Try to find examples of both possessives and contractions. Have the apostrophes been used correctly? Rewrite any that are incorrect.

Grammar

As with spelling and punctuation, there are a great many grammar rules. For practical purposes, however, you really need to know relatively few. This section focuses only on the basics—the rules governing sentence fragments, run-on sentences, and agreement.

Sentence Fragments

As the term itself denotes, a sentence fragment is an incomplete sentence. Most fragments are actually the result of faulty punctuation—

when a writer inserts end punctuation too soon, thereby "stranding" part of the sentence. Consider these examples:

> Rabies has been a problem since the 1950s throughout New York
>
> (Fragment)
>
> State. Including Long Island and New York City.
>
> [The first period should have been a comma.]

> (Fragment)
>
> If you cut yourself while skinning an animal. Have your local health agency check the animal for rabies.
>
> [Again, the first period should have been a comma.]

You can usually avoid sentence fragments if you remember three basic principles:

1. To be complete, a sentence must include a subject (actor) and a verb (action).

   (Subject) (Verb)

   Snowmobilers sometimes take unnecessary risks.

2. If a sentence begins with a word or phrase that seems to point toward a two-part idea (for example, "**if** this, then that."), the second part must be included within the sentence, because the first part is a fragment and therefore cannot stand alone.

   Here are some examples of words that signal a two-part idea:

after	if
although	since
because	unless
before	until
for	when

3. Certain kinds of verb forms (-ed forms, -ing forms, and "to" forms) cannot serve as the main verb in a sentence.

   (Fragment)

   Opened in 1939. The Merritt Parkway in Connecticut was one of America's first freeways.

   (Correct Sentence)

   Opened in 1939, the Merritt Parkway in Connecticut was one of America's first freeways.

(Fragment)

Rising to a height of 1454 feet. The Sears Tower in Chicago replaced the World Trade Center as the tallest skyscraper in the United States.

(Correct Sentence)

Rising to a height of 1454 feet, the Sears Tower in Chicago replaced the World Trade Center as the tallest skyscraper in the United States.

(Fragment)

To succeed in your own business. You need both energy and luck.

(Correct Sentence)

To succeed in your own business, you need both energy and luck.

Notice that the -ed, -ing, and "to" forms frequently appear in introductory phrases. Learn to recognize these phrases for what they are—not sentences in themselves, but beginnings of sentences—and punctuate each with a comma, not a period. (See rule number two in the section on commas.)

Run-On Sentences

While the sentence fragment is something to avoid, even worse is a sentence that goes on and on after it should have stopped. A run-on sentence spills over into the following sentence with no break in between. When that happens, the writing takes on a rushed, headlong quality, and ideas become jumbled together.

There are two ways that a sentence can overflow into the next: either with a comma weakly separating the two sentences, or with nothing at all in between. Here is an example of each:

A surveyor's measurements must be precise, there is no room for error.

A surveyor's measurements must be precise there is no room for error.

Technically, only the second of these examples is a true run-on. The first is really an instance of what grammarians refer to as a "comma splice." For practical purposes, however, the problem is the same. In both cases, the first sentence has collided with the second. Obviously, the two sentences must be separated with a period.

A surveyor's measurements must be precise. There is no room for error.

Another option would be to use a comma along with a linking word to join the two sentences.

> A surveyor's measurements must be precise, as there is no room for error.

Or you may prefer to turn one of the sentences into a fragment and use it as an introductory construction.

> Since there is no room for error, a surveyor's measurements must be precise.

It should be clear by now that fragments and run-ons alike are usually the result of faulty punctuation. Certain patterns are correct, while others are not, as the following chart indicates:

Correct Patterns	**Incorrect Patterns**
sentence.	fragment.
sentence. sentence.	fragment. fragment.
sentence, link sentence.	sentence, sentence.
fragment, sentence.	fragment. sentence.
sentence, fragment.	sentence. fragment.
fragment, sentence, fragment.	fragment. sentence. fragment.

Subject–Verb Agreement

Another common grammar error is to use a plural verb with a singular subject, or vice-versa. Remember that a singular subject requires a singular verb, while a plural subject requires a plural verb.

(Singular Subject) (Singular Verb)
A welder welds.

(Plural Subject) (Plural Verb)
Welders weld.

Note that singular subjects rarely end in -s, while singular verbs usually do. Conversely, plural subjects usually do end in -s, but plural verbs never do.

Although the subject–verb agreement rules may seem obvious, many writers commit agreement errors simply because they fail to distinguish between singular and plural subjects. This sometimes occurs when the subject is an "indefinite pronoun," most of which are singular. Here is a chart of the most common indefinite pronouns, indicating which ones are singular, which plural, and which can function as either.

Singular			**Plural**	**Either**
anybody	everybody	no one	few	all
anyone	everyone	nothing	many	any
anything	everything	somebody	several	more
each	neither	someone		most
either	nobody	something		none
				some

Agreement errors can also result when there is a cluster of words between the subject and its verb, thereby creating a misleading sound pattern.

INCORRECT A pile of tools are on the workbench.

CORRECT A pile of tools is on the workbench.

Even though "tools are" sounds correct, the first sentence is incorrect because "pile"—not "tools"—is the (singular) subject, and therefore requires the singular verb "is."

Pronoun–Antecedent Agreement

Just as subjects and verbs must agree, so too must pronouns and their antecedents (the words that the pronouns refer back to).

(Singular Antecedent) (Plural Pronoun)

For a woman to succeed as an umpire, they must overcome much prejudice.

(Singular Antecedent) (Singular Pronoun)

For a woman to succeed as an umpire, she must overcome much prejudice.

The first sentence is incorrect because "a woman," which is singular, disagrees with "they," which is plural. (Although "they" is often used as a singular in speech, it must always be treated as a plural in writing.) The second sentence is correct because both "a woman" and "she" are singular, and therefore agree.

Once again, indefinite pronouns can create agreement problems.

(Singular Antecedent)
Everyone on the men's basketball team should be proud of
(Plural Pronoun)
themselves.

(Singular Antecedent)
Everyone on the men's basketball team should be proud of
(Singular Pronoun)
himself.

The first sentence is incorrect because "everyone," which is singular, disagrees with "themselves," which is plural. The second sentence is correct because both "everyone" and "he" are singular, and therefore agree.

Let us consider one more aspect of agreement, using this sentence as a starting point:

(Singular Antecedent) (Plural Pronoun)
Everybody should mind their own business.

Clearly, there is disagreement between "Everybody," which is singular, and "their," which is plural, even though this is how the sentence would probably be worded in speech. Since writing is more formal than speech, the problem must be corrected. There are two ways to do so: either the pronoun and its antecedent can both be plural, or both can be singular.

Here are three different singular versions:

(Singular Antecedent) (Singular Pronoun)
Everybody should mind his own business.

(Singular Antecedent) (Singular Pronoun)
Everybody should mind her own business.

(Singular Antecedent) (Singular Pronoun)
Everybody should mind his or her own business.

Here are two plural versions:

(Plural Antecedent) (Plural Pronoun)
People should mind their own business.

(Plural Antecedent) (Plural Pronoun)
We should all mind our own business.

The plural approach is almost always better because it enables us to avoid gender-biased language without resorting to wordy "his or her" constructions. For a more in-depth treatment of this topic, see Appendix A, pages 276–278.

EXERCISE B.9

In both of the following passages, nearly all punctuation and capital letters are missing. Copy the passages, inserting punctuation and capitals to prevent fragments and run-ons.

1. In many parts of the United States building codes had long ignored the severity of earthquakes and the damage they can cause nevertheless research has revealed that earthquakes and earth tremors are a threat in a far greater percentage of the country than most people commonly believe while documentation shows that earthquakes have not resulted in loss of life or costly damage along the east coast cataloging of this kind of information has been sporadic and inconsistent it was not until 1900 that instruments were used to record and locate this activity ever since then information has been gathered and more uniformly recorded by several organizations designers of building codes nationwide have used this information to define requirements for withstanding the lateral forces of earthquakes these forces are generated from the movement of the ground due to seismic activity and converted into horizontal and vertical loads building frames architectural components and mechanical and other equipment must be designed to withstand these forces in areas where quakes have occurred more recently and cause extensive damage and loss of life these design parameters have long been incorporated into structural design elsewhere however the building codes are just beginning to reflect the potential damage of seismic activity.
2. One of the fears associated with electric cars is that the batteries will run down and leave you stranded this is not the same as run-

ning out of gasoline because you cannot bring electricity back in a can or make a quick stop at a service station when you're empty even an on-board battery charger won't do any good by the side of the road because such recharging requires time and a source of electricity within reach of an extension cord one way to eliminate this fear is to avoid electric cars the general public has taken this course ever since electric self-starters were added to gas buggies in 1912 however the oil embargoes in the mid-1970s created such hysteria over the thought of gas rationing that electric vehicles became plausible once again even crude electric cars that were little more than golf carts with headlights and turn signals were finding buyers that fear of being stranded still existed though and one answer to the problem was a hybrid gasoline/electric car it was in this atmosphere that Briggs & Stratton commissioned a hybrid electric car using conventional batteries in conjunction with one of its larger gas engines.

EXERCISE B.10

All but one of the following sentences contain agreement errors. Rewrite the sentences to correct the errors.

1. Regularly scheduled sessions of hypnotherapy creates a state of complete mental and physical relaxation.
2. Each of the players on the women's basketball team will receive their trophy next week.
3. Although secretaries are often underpaid, their role are essential to any workplace.
4. Neither of the security guards were in the plant at the time of the break-in.
5. Sophisticated harvesting machines like the one shown in Figure 3 is in use on many farms today.
6. A student must study hard if he or she expect to succeed in college.
7. Classes in soil science provides future engineers with experience in testing various types of soil to determine its characteristics.
8. There is more than two million miles of paved roads in the United States.
9. Everyone in the apartment building will have their rent raised next month.
10. Either Smith or Jones are guilty.

Index